Water-in-Plants Bibliography

volume 9 1983

References no. 11193–12911 / ABD–ZUR

Editors J. Pospíšilová and J. Solárová

1984 **DR W. JUNK PUBLISHERS**
a member of the KLUWER ACADEMIC PUBLISHERS GROUP
DORDRECHT / BOSTON / LANCASTER

Contributors

J. Solárová
J. Pospíšilová
J. Čatský
Z. Šesták
I. Tichá

ISBN 978-90-6193-520-9 ISBN 978-94-009-5510-3 (eBook)
DOI 10.1007/978-94-009-5510-3

PREFACE

The ninth volume of Water-in-Plants Bibliography includes papers in all fields of plant water relations research which appeared during the year 1983. - from theoretical considerations about the state of water in cells and its membrane transport to drought resistance of plants or physiological significance of irrigation. In addition to papers devoted entirely to plant water relations, papers on other topics are included if they contain data on plant hydration level, water vapour efflux, rate of water uptake or water transport, etc., or if they contain valuable methodological information (measurement of selected microclimatic factors, soil moisture etc.).

We have tried to cover fully the relevant papers which have been published in important scientific periodicals and books. Articles appeared in local journals, mimeographed booklets, abstracts of thesis and of symposia contributions, etc., were chosen mostly from reprints received directly from authors. The courtesy of those is highly appreciated. The manuscript is usually prepared in May and June of the year following the year which it covers. Unfortunately some reprints come later and thus the respective references appear in the following volume, with one year delay.

To maximize the value of the bibliography the references are arranged alphabetically according to the authors' names, and each volume is provided with three indexes. The authors' index contains all names of authors, co-authors and editors. Plant genera used as experimental material are indexed according to their Latin names. The subject index covers primary items chosen according to the interest of water relations researchers. Its preparation was based not only on the titles, key words and abstracts but also on the whole content of the article. By combining two or more items a more detailed information may be obtained.

Since more than 1500 relevant papers dealing with plant water relations and relative topics are published every year and included in this bibliography, and since all citations have been checked with the originals, collecting and preparing for publication such a large amount of material would have been impossible without the collaboration of our colleagues from the Department of Physiology of Photosynthesis and Water Relations of the Institute of Experimental Botany of the Czechoslovak Academy of Sciences in Prague. We have also acknowledge with thanks the cooperation of Mrs. Ludmila Hávová and Mrs. Jindřiška Srbová who helped in typing card material and the librarian of our Institute Mrs. Zora Zawoyská who helped us with checking the references.

Dr. Jana Pospíšilová and Ing. Jarmila Solárová

Institute of Experimental Botany
Czechoslovak Academy of Sciences

Flemingovo nám. 2
160 00 PRAHA 6
Czechoslovakia

Praha, 8 October 1984

INSTRUCTIONS FOR USE

All references are arranged alphabetically according to the authors' names. They are numbered and these numbers are used in the indexes. An asterisk preceding the number denotes the reference published in the preceding period (1975 - 1981).

Authors' names are presented in the spelling used in the original paper. If this spelling does not correspond to the spelling usually used by the author (e.g. Russian papers of English authors), one spelling is referred to the other in the Authors' index. Like the transcriptions they are alphabetically arranged mostly according to the authors' own references. Nevertheless, the editors apologize for some misinterpretations which are partly corrected by the cross-indexing in the Authors' index.

The references contain the original unshortened title of the paper (book). English, French, and German titles are cited in the original language. Titles in other languages are supplemented with a translation in English (using the title of the respective English abstract, if it is presented). Titles of Japanese, Chinese etc. papers are given in English translation only. In both these cases the abbreviations of the original language and the language of the abstract are given in brackets at the end of the reference. The following abbreviations are used most frequently:

Belorussian	**Jap**anese
Bulgarian	**Latv**ian
Chinene	**Lithu**anian
Croatian	**Norweg**ian
Danish	**Pol**ish
English	**R**ussian
Esthonian	**Roum**anian
French	**Slov**ak
German	**Span**ish
Georgian	**Swed**ish
Hungarian	**Ukr**ainian
Italian	**Uzb**eg

The transliteration of Cyrillic characters is in accordance with the BSI-ASA/SC-Z39 draft table, i.e.:

a	а	p	п
b	б	r	р
ch	ч	s	с
d	д	sh	ш
e	е	shch	щ
ė	э	t	т
f	ф	ts	ц
g	г	u	у
i	и	v	в
ĭ	й	y	ы
k	к	ya	я
kh	х	yu	ю
l	л	z	з
m	м	zh	ж
n	н	"	ъ
o	о	'	ь

Several exceptions apply for Ukrainian and Belorussian:

Ukrainian:	y	и
	i	i
	ĭ	ї
Belorussian:	ŭ	ў

The journals' names are abbreviated mainly according to the Style Manual for Biological Journals (Second Edition, American Institute of Biological Sciences, Washington, D.C. 1964), e.g.:

Abhandlungen
Abstract
Abteilung
Academy
Acker
Acta
Advances
Africa (-ican)
agricultural
Agriculture
Agrobiology (-ogiya)
Agrobotanica
Agrokémia
Agronomy
agropecuaria
Akademie (-emiya)
Algology
allgemeine
Amélioration
America
American
Anais (-alele)
Analysis
analytical
Anatomy
angewante
animal
Annales (-als)
annual
anorganic (-anisch)
applied
aquatic
Arbeit
Archiv
Argentina
Association
Atmosphere
atmospheric
atomic
Australia (-ralian)
Azerbaĭdzhanskaya
Bacteriology
Beiheft
Beiträge
Belgique
Belorusskaya
Berichte
biochemical
Biochemie
Biochemistry
biochimica
biokhimicheskiĭ
Biokhimiya
Bioklimatologie
Biologia (-ogy)
biological (-ogisk)
biophysical
Biophysics
Bodenkunde
Boletin (-ettino)
Bolgarskiĭ
botanica (-anicorum)
botanical (-anisca)
Botanika (-any)
Brasileira
Brazil
Breeding
British
Bulletin (-etins)
Byulleten
California
Canada (-adian)
cellular (-ulaire)
Center
central
Centralblatt
Československý
chemical
Chemistry
chimicus
Chinese
Chromatography
Chronicle
Ciencia
cientificas
College
Commision
Communication
comparative
Comptes Rendus
Conference
Congress
Conservation
Contamination
Contribution
Control
Croatica
cultural
Culture
current
Cytobiology
Cytochemistry
Cytology
Czechoslovak
Danske
dendrological
Dendrology
Department
Deutsche (-schland)
Development
Disease
Dissertation
Division
Doklady
Dopovidi
Drainage
ecological
Ecology
Economy
Edafology
Education
Ėkologiya
ėksperimental'nyĭ
Embryology
Encyclopedia
Engineering
Enology
Entomology
environmental
Enzymology
Estonskaya
European
Experiment
experimental
Faculty
Federation
Fizika
Fiziologiya
Flurbereinigung
forestiere
Forestry
Forschung
Foundation
France
Gazette
general
genetical
geneticheskiĭ
Genetics (-ika)
Geobotany
Geofizika
Geophysics
Gesellschaft
Giornale
gosudarstvennyĭ
Government
Grassland
Gruzinskaya
Helveticus
Histochemistry
Histoire (-ory)
Histology
horticultural
Horticulture
Hungaricae
Hungaricus
Husbandry
Hydrobiology
Hydrology
Indian
Industry
inorganic
Institute
Institutului
international
Investigation
Irrigation
Isotopes
issledovatel'skiĭ
Italian (-y)
Izvestiya
Jahrbuch
Japan (-anese)
Journal
Khimiya
Klasse
Kongelige
Közlemenyek
kul'turnykh
Laboratory
Landbauforschung
Landwirtschaft
lesní (-ího)
Letters
Limnology
Linnean
Litovskoĭ
Lucrarile
Magazin
Management
marina (-ine)
Material
Mathematics
Mededelingen
mediterranean
Meldinger
Meteorology
Microbiology
Midland
Mitteillungen
Modeling
modern
molecular
Monographiae (-aphy)
Moskovskiĭ (-ovskogo)
Mycology
national
natural
Naturalist
naturelle
naturkundliche
Naturforschung
nauchnye (-nyĭ)
Neerlandica
Netherland
New Zealand
Norges
Norwegian
Notiser
nuclear
Nutrition
obshcheĭ (-iĭ)
Oceanography
Oecologia
Ökologie
Optics
opytnaya (-yĭ)
organic
original
ornamental
Otdelenie
Paleobotany
Palynology
Pathology
pedagogicheskiĭ
Pesquisa
Pesticide
Pflanzen-
Pflanzenernährung
Pflanzenphysiology
Pflanzenzüchtung
Philosophy
Photogrammetric
Phycology
physical
Physics

physiological
Physiology
Phytologist
Phytopathology
Phytotaxonomy
Plantarum
Polonica (-ska)
Pollution
Práce
Practice
prikladnoĭ
Proceedings
Progress
Publishers
Quality
quantitative
Quarterly
Radiation
Radiobiology
Rasteniĭ
Rastenievodstvo
Recherche (-erches)
Report
Research
Resources
Review (-ista, -ue)
Rivista
Roczniky
rolniczych
Rostlin (-lina)
rostlinná
Roumaine
royal
Russian
Russkiĭ
Sborník
Scandinavica
Scandinavicus
School
Science
scientific
Section
Selskabs
Sel'skokhozyaĭstvo
Series (-iya)
Service
Shkola (-oly)
Sibirskiĭ (-skogo)
Skrifter
Slovak (-enská)
Society
Soobshcheniya
Sovetskiĭ (-iet)
sovremennyĭ
special
sperimentale
SSSR
Stantsii (-ntsiya)
Statio
stiintifice
subtropical (-icale)
summary
Supplement
Survey
Swedish
Symposium
System
Tagungsberichte
technical (-nische)
Technology
theoretical
thermal
Tidsskrift
Tijdschrift
Toxicology
Transactions
Travail (-aux)
tropical (-icale)
Trudy
Turkmenskaya
uchenye
Ugeskrift
United Kingdom
Ukrainian
Ukrains'kaya
Universidad (-ersity)
US. USA
USSR
Uzbekskiĭ (-ekskaya)
vědecké (-ecký)
vegetable
végétale
Verhandlungen
Veröffentlichungen
Vestnik
Videnskabernes
Virology
Virusforschungen
Volume
Voprosy
vostochnyĭ
vsesoyuznyĭ
vyssheĭ (-iĭ)
výzkumný (-umného)
Weekblatt
Wetenschappen
Wissenschaft
Zapiski
Zeitschrift
Zeitung
Zentralblatt
Zhurnal

The numbers at the end of each reference of a journal article denote: volume (issue) : first page - last page, year of publication. The number of issue is given only in journals where each issue is paginated separately.

Book titles are cited according to the title page, not to the book jacket or cover. The publishing house, place and year of publication are included.

Printers' errors in the original papers are marked by underlining the respective words (letters).

*11193 - **ABD EL-HAFEEZ, A.A.:** Comparative ET studies on Berseem clover grown at different water table depths. I. Under rainfall conditions. - Acta agron. Acad. Sci. Hung. 30: 415-424, 1981.

11194 - **ABDELLY, C., SELLAMI, A.:** Variation de la taille du pool des nucléotides adényliques chez des halophytes, en relation avec l'état hydrique du sol. - C.R. Acad. Sci. (Paris) Sér. III 297: 91-93, 1983.

11195 - **ABDUL-JABBAR, A.S., SAMMIS, T.W., LUGG, D.G., KALLSEN, C.E., SMEAL, D.:** Water use by alfalfa, maize, and barley as influenced by available soil water. - Agr. Water Manage. 6: 351-363, 1983.

*11196 - **ABDULLAEV, Kh.A., KASPAROVA, I.S., NASYROV, Yu.S.:** Vliyanie solevogo stressa na intensivnost' fotosinteza i fotodykhanie khlopchatnika. [Effect of salt stress on photosynthetic rate and photorespiration of cotton.] - Dokl. Akad. Nauk Tadzh. SSR 25: 306-308, 1982. [In R, ab: Tadj.]

11197 - **ABOD, S.A., SANDI, S.:** Effect of restricted watering and its combination with root pruning on root growth capacity, water status and food reserves of *Pinus caribaea* var. *hondurensis* seedlings. - Plant Soil 71: 123-129, 1983.

11198 - **ABO-HAMED, S., COLLIN, H.A., HARDWICK, K.:** Biochemical and physiological aspects of leaf development in cocoa (*Theobroma cacao* L.). VII. Growth, orientation, surface structure and water loss from developing flush leaves. - New Phytol. 95: 9-17, 1983.

11199 - **ACEVEDO, E., BADILLA, I., NOBEL, P.S.:** Water relations, diurnal acidity changes, and productivity of a cultivated cactus, *Opuntia ficus-indica*. - Plant Physiol. 72: 775-780, 1983.

*11200 - **ACHARYA, C.L., SANDHU, S.S., ABROL, I.P.:** Note on the effect of subsoil sodicity on the water use by irrigated wheat. - Ind. J. agr. Sci. 52: 872-873, 1982.

11201 - **ACKERMAN, W.L., ZHANG, A.-L.:** Distinguishing *Camellia* L. species using dorsal leaf surface impressions. - J. Amer. Soc. hort. Sci. 108: 439-444, 1983.

11202 - **ACKERSON, R.C.:** Comparative physiology and water relations of two corn hybrids during water stress. - Crop Sci. 23: 278-283, 1983.

11203 - **ACKERSON, R.C., RADIN, J.W.:** Abscisic acid accumulation in cotton leaves in response to dehydration at high pressure. - Plant Physiol. 71: 432-433, 1983.

11204 - **ADAMS, C.A., FJERSTAD, M.C., RINNE, R.W.:** Characteristics of soybean seed maturation: necessity for slow dehydration. - Crop Sci. 23: 265-267, 1983.

*11205 - **ADARA, O.A., KANNENBERG, L.W.:** Performance under stress of advanced cycles of four populations of corn (*Zea mays* L.) in an S_1 *per se*. Recurrent selection program. - Can. J. Plant Sci. 61: 29-36, 1981.

11206 - **ADDICOTT, F.T. (ed.):** Abscisic Acid. - Praeger Publishers, New York 1983.

11207 - **AHLRICHS, J.S., BAUER, M.E.:** Relation of agronomic and multispectral reflectance characteristics of spring wheat canopies. - Agron. J. 75: 987-993, 1983.

*11208 - **AHMAD, I., WAINWRIGHT, S.J., STEWART, G.R.:** The solute and water relations of *Agrostis stolonifera* ecotypes differing in their salt tolerance. - New Phytol. 87: 615-629, 1981.

*11209 - **AHMADI, M.S., HADERLIE, L.C., WICKS, G.A.:** Effect of growth stage and water stress on barnyardgrass (*Echinochloa crus-galli*) control and on glyphosate absorption and translocation. - Weed Sci. 28: 277-282, 1980.

11210 - **AHMADI, N.**: Variabilité génétique et hérédité de mécanismes de tolérance à la sécheresse chez le riz *Oryza sativa* L.: I. Développement du système racinaire. - Agron. trop. 38: 110-117, 1983.

11211 - **AHMADI, N.**: Variabilité génétique et hérédité de mécanismes de tolérance à la sécheresse chez le riz *Oryza sativa* L. II. Sensibilité stomatique aux déficits hydriques. - Agron. trop. 38: 118-122, 1983.

11212 - **AHMED, A.M., HEIKAL, M.D., HINDAWY, O.S.**: Side effects of Benomyl (fungicide) treatments on sunflower, cotton and cowpea plants. - Phyton 23: 185-195, 1983.

11213 - **AHMED, A.M., HEIKAL, M.D., SHADDAD, M.A.**: Changes in growth, photosynthesis and fat content of some oil producing plants over a range of salinity stresses. - Acta agron. Acad. Sci. Hung. 32: 370-375, 1983.

*11214 - **AHMED, A.M., HEIKAL, M.M., ZIDAN, M.A.**: Photosynthesis and some other growth parameters of some leguminous plants as affected by salinization treatments. - Natur. Montspeliensia, Sér. Bot. 33: 1-10, 1980.

*11215 - **AINA, P.O.**: Influence of soil moisture, soil compaction and K fertilization on maize (*Zea mays* L.) grown on Iwo Soil. - Z. Pflanzenernaehr. Bodenk. 143: 317-323, 1980.

11216 - **AKEY, W.C., MORRISON, I.N.**: Effect of moisture stress on wild oat (*Avena fatua*) response to diclofop. - Weed Sci. 31: 247-253, 1983.

*11217 - **AKIMOVA, T.V., KRITENKO, S.P.**: Vliyanie zakalivayushchikh i povrezhdayushchikh temperatur na nekotorye pokazateli vodnogo i ènergeticheskogo obmena v list'yakh ogurtsa. [Effect of hardenning and damaging temperatures on some characteristics of water and energy balance in cucumber leaves.] - In: Vliyanie Faktorov Vneshneĭ Sredy i Fiziologicheski Aktivnykh Veshchestv na Termorezistentnost' i Produktivnost' Rasteniĭ. Pp. 43-50. Karel'skiĭ Filial Akad. Nauk SSSR, Petrozavodsk 1982. [In R.]

11218 - **ALAERTS, M., BADJI, M., FEYEN, J.**: Effect of the simulation procedures of root water uptake on the assesment of the plant water stress. - In: MARCELLE, R., CLIJSTERS, H., VAN POUCKE, M. (ed.): Effects of Stress on Photosynthesis. Pp. 153. Martinus Nijhoff / Dr W. Junk Publishers, The Hague - Boston - London 1983.

*11219 - **ALBRIGO, L.G., LYRENE, P.M., FREEMAN, B.**: Waxes and other surface characteristics of fruit and leaves of native *Vaccinium elliotti* Champ. - J. Amer. Soc. hort. Sci. 105: 230-235, 1980.

*11220 - **ALDIS, D.F., FOSTER, G.H.**: Moisture changes in grain from exposure to ambient air. - Trans. ASAE 23: 753-760, 1980.

11221 - **ALLEN, E.B., CUNNINGHAM, G.L.**: Effects of vesicular-arbuscular mycorrhizae on *Distichlis spicata* under three salinity levels. - New Phytol. 93: 227-236, 1983.

11222 - **ALLEN, M.F., BOOSALIS, M.G.**: Effects of two species of VA mycorrhizal fungi on drought tolerance of winter wheat. - New Phytol. 93: 67-76, 1983.

*11223 - **ALMADI, L.**: Vízháztartási vizsgálatok I. [Study of water regime I.] - Bot. Közlem. 69: 85-93, 1982. [In Hung, ab: E.]

11224 - **AL-NAKSHABANDI, G.A.**: The potential evapotranspiration of Short grass (*Cynodon dactylon*) as related to the estimated potential evaporation and evapotranspiration from meteorological data in the coastal region of Kuwait. - J. Arid Environ. 6: 33-38, 1983.

*11225 - **ALONI, B., PRESSMAN, E.**: Interaction with salinity of GA_3-induced leaf elongation, petiole pithiness and bolting in celery. - Scientia Hort. 13: 135-142, 1980.

*11226 - **AL-RASHIDI, R.K., LOYNACHAN, T.E., FREDERICK, L.R.**: Desiccation tolerance of four strains of *Rhizobium japonicum*. - Soil Biol. Biochem. 14: 489-493, 1982.

*11227 - **ALVINO, A., FRUSCIANTE, L., MONTI, L.M.**: Yield and quality traits of two new tomato varieties for peeling under different irrigation regimes. - Acta Hort. 100: 173-180, 1980.

11228 - **ALVINO, A., ZERBI, G., FRUSCIANTE, L., MONTI, L.M.**: Evaluation of field bean lines grown with a shallow water table maintained at different levels. - Field Crop Res. 6: 179-188, 1983.

*11229 - **AMBROS, Z.**: Produkčný potenciál lesných ekosystémov. [Production potential of forest ecosystems.] - Lesn. Časopis (Bratislava) 28: 315-326, 1982. [In Slov, ab: E, G, R.]

11230 - **AMBROS, Z.**: Hydrický potenciál lesných ekosystémov. [Hydric potential of forest ecosystems.] - Lesn. Časopis (Bratislava) 29: 3-13, 1983. [In Slov, ab: E, G, R.]

11231 - **AMTHOR, J.S.**: Sorghum seedling growth as a function of sodium chloride salinity and seed size. - Ann. Bot. 52: 915-917, 1983.

11232 - **ANDERSON, J.A., GUSTA, L.V., BUCHANAN, D.W., BURKE, M.J.**: Freezing of water in citrus leaves. - J. Amer. Soc. hort. Sci. 108: 397-400, 1983.

*11233 - **ANDONOV, D.**: Ikonomicheska efektivnost ot tsvekloproizvodstvoto pri napoyavane chrez d"zhduvane v usloviyata na Yambol. [Economic effectiveness of sugar beet production under irrigation in Yambol.] - Rasteniev. Nauki 19(8): 42-48, 1982. [In Bulg, ab: R, E.]

11234 - **ANDONOVA, T., ANDONOVA, P.**: Kolichestveni izmeneniya na svobodnite tsitokinini v mladi tsarevichni rasteniya pri pochveno zasushavane. [Changes in the free cytokinins of maize seedlings during soil drought.] - Fiziol. Rast. (Sofia) 9(1): 55-59, 1983. [In Bulg, ab: R, E.]

*11235 - **ANGUS, D.E.**: Effect of soil moisture on evapotranspiration. - In: Agricultural Engineering Conference. Pp. 184-187. Geelong 1980.

11236 - **ANGUS, J.F., HASEGAWA, S., HSIAO, T.C., LIBOON, S.P., ZANDSTRA, H.G.**: The water balance of post-monsoonal dryland crops. - J. agr. Sci. 101: 699-710, 1983.

*11237 - **ANGUS, J.F., ZANDSTRA, H.G.**: Climatic factors and the modeling of rice growth and yield. - In: Agrometeorology of the Rice Crop. Pp. 189-199. World Meteorological Organization, International Rice Research Institute, Los Baños 1980.

*11238 - **ANTIPOV, N.I.**: Osobennosti vodoobmena rozlichnykh tipov plodov rasteniĭ. [Characteristics of water exchange in different types of fruits.] - Ėkologiya 1980(2): 86-88, 1980. [In R.]

11239 - **ANTLFINGER, A.E., DUNN, E.L.**: Water use and salt balance in three salt marsh succulents. - Amer. J. Bot. 70: 561-567, 1983.

*11240 - **ANTOSZEWSKI, R., LENZ, F.**: Regulation of transport between leaf and fruit. - In: Proceedings 21st International Horticultural Congress. Volume I. Pp. 91-106. International Society for Horticultural Science, Wageningen 1982.

11241 - **APARICIO-TEJO, P.M., BOYER, J.S.:** Magnesium nutrition and grain yield of maize having low water potential. - Agron. J. 75: 919-923, 1983.

11242 - **APPLEBY, R.F., DAVIES, W.J.:** A possible evaporation site in the guard cell wall and the influence of leaf structure on the humidity response by stomata of woody plants. - Oecologia 56: 30-40, 1983.

11243 - **APPLEBY, R.F., DAVIES, W.J.:** The structure and orientation of guard cells in plants showing stomatal responses to changing vapour pressure difference. - Ann. Bot. 52: 459-468, 1983.

11244 - **ARAD (MALIS), S., MIZRAHI, Y.:** Stress-induced ripening of the non-ripening tomato mutant *nor*. - Physiol. Plant. 59: 213-217, 1983.

11245 - **ARAŨJO, J.P., de, FREIRE, L.C., MAGALHÃES, A.A., de, FARIA, C.M.B., de:** Sistema de produção para culturas irrigadas no submédio São Francisco. I. Tomate industrial. [Farming system for irrigated crops in the São Francisco region. I. Industrial tomato.] - Pesq. agropec. Bras. 18: 1-3, 1983. [In Port, ab: E.]

11246 - **ARGEL, P.J., HUMPHREYS, L.R.:** Environmental effects on seed development and hardseededness in *Stylosanthes hamata* cv. Verano. II. Moisture supply and illuminance. - Aust. J. agr. Res. 34: 271-277, 1983.

11247 - **ARGEL, P.J., HUMPHREYS, L.R.:** Environmental effects on seed development and hardseededness in *Stylosanthes hamata* cv. Verano. III. Storage humidity and seed characteristics. - Aust. J. agr. Res. 34: 279-287, 1983.

11248 - **ARMITAGE, A.M., KOWALSKI, T.:** Effect of irrigation frequency during greenhouse production on the postproduction quality of *Petunia hybrida* Vilm. - J. Amer. Soc. hort. Sci. 108: 118-121, 1983.

11249 - **ARMITAGE, A.M., VINES, H.M., TU, Z.-P., BLACK, C.C.:** Water relations and net photosynthesis in hybrid geranium. - J. Amer. Soc. hort. Sci. 108: 310-314, 1983.

*11250 - **ARTECA, R.N.:** Effect of root applications of kinetin and gibberellic acid on transplanting shock in tomato plants. - HortScience 17: 633-634, 1982.

11251 - **ASFARY, A.F., WILD, A., HARRIS, P.M.:** Growth, mineral nutrition and water use by potato crops. - J. agr. Sci. 100: 87-101, 1983.

*11252 - **ASHLEY, D.A.:** Soybean. - In: TEARE, I.D., PEET, M.M. (ed.): Crop-Water Relations. Pp. 389-422. John Wiley & Sons, New York - Chichester - Brisbane - Toronto - Singapore 1982.

11253 - **ASHWORTH, E.N., ROWSE, D.J., BILLMYER, L.A.:** The freezing of water in woody tissues of apricot and peach and the relationship to freezing injury. - J. Amer. Soc. hort. Sci. 108: 299-303, 1983.

11254 - **ASPINALL, D., PARAMESWARAN, K.V.M., GRAHAM, R.D.:** Proline accumulation in the grains, floral organs and flag leaves of wheat and barley in response to variations in water and nitrogen supply. - Irrig. Sci. 4: 157-166, 1983.

11255 - **ASTADZHOV, N.:** Vliyanie na polivaneto v"rkhu razvitieto i produktivnostta na blatnoto kokiche. [The effect of irrigation on the development and produktivity of the marsh snowflake.] - Rasteniev. Nauki 20(1): 50-56, 1983. [In Bulg, ab: R, E.]

11256 - **ATHERTON, J.G., FAROOQUE, A.M.:** High temperature and germination in spinach. II. Effects of osmotic priming. - Scientia Hort. 19: 221-227, 1983.

11257 - **ATTIA, A., JEANJEAN, R.:** Influence of osmotic shock and of plasmolysis on phosphate uptake by excised corn roots. - Physiol. vég. 21: 36-47, 1983.

*11258 - **ATTIWILL, P.M., CROMER, R.N.:** Photosynthesis and transpiration of *Pinus radiata* D. Don under plantation conditions in southern Australia. I. Response to irrigation with waste water. - Aust. J. Plant Physiol. 9: 749-760, 1982.

*11259 - **ATTIWILL, P.M., SQUIRE, R.O., NEALES, T.F.:** Photosynthesis and transpiration of *Pinus radiata* D. Don under plantation conditions in southern Australia. II. First-year seedlings and 5-year-old trees on aeolian sands at Rennick (south-western Victoria). - Aust. J. Plant Physiol. 9: 761-771, 1982.

11260 - **AUGUSTIN, B.J., KARNOK, K.J.:** A controlled environment system for turfgrass research. - Agron. J. 75: 306-310, 1983.

11261 - **AUJLA, T.S., CHEEMA, S.S.:** Modifying profile water storage through tillage, herbicide, chemical evaporation retardent and straw mulch and its effect on rainfed chickpea (*Cicer arietinum* L.). - Soil Tillage Res. 3: 159-170, 1983.

11262 - **AULD, B.A., KEMP, D.R., MEDD, R.W.:** The influence of spatial arrangement on grain yield of wheat. - Aust. J. agr. Res. 34: 99-108, 1983.

11263 - **AUSSENAC, G., FINKELSTEIN, D.:** Influence de la sécheresse sur la croissance et la photosynthèse du cèdre. - Ann. Sci. forest. 40: 67-77, 1983.

11264 - **AUSSENAC, G., LEVY, G.:** Influence du dessèchement du sol sur le comportement hydrique et la croissance du chêne pédonculé (*Quercus pedunculata* Ehrl.) et du frêne (*Fraxinus excelsior* L.) cultivés en cases de végétation. - Ann. Sci. forest. 40: 251-264, 1983.

11265 - **AZAM-ALI, S.N.:** Seasonal estimates of transpiration from a millet crop using a porometer. - Agr. Meteorol. 30: 13-24, 1983.

11266 - **AZCÓN-BIETO, J.:** The effect of carbohydrate status on the photosynthetic, stomatal and respiratory physiology of wheat leaves. - In: Sixth International Congress on Photosynthesis. Abstracts. Volume 2. Pp. 121. Brussels 1983.

11267 - **BADANOVA, K.A., PRUSAKOVA, L.D.:** Povyshenie zasukhoustoǐchivosti yarovykh pshenits i yachmenya metodom predposevnogo zakalivaniya i obrabotki retardantami. [Increase in drought resistance of spring wheat and barley by method of presowing hardening and treatment with herbicides.] - Sel'skokhoz. Biol. 1983(3): 41-44, 1983. [In R, ab: E.]

11268 - **BADJI, M., ALAERTS, M., FEYEN, J.:** Simulation of water stress in *Brachiaria ruziziensis* under a high evaporative demand. - In: MARCELLE, R., CLIJSTERS, H., VAN POUCKE, M. (ed.): Effects of Stress on Photosynthesis. Pp. 153. Martinus Nijhoff / Dr W. Junk Publishers, The Hague - Boston - London 1983.

11269 - **BADZINSKI BUHL, M., STEWART, C.R.:** Effects of NaCl on proline synthesis and utilization in excised barley leaves. - Plant Physiol. 72: 664-667, 1983.

11270 - **BAENZIGER, P.S., WESENBERG, D.M., SICHER, R.C.:** The effects of genes controlling barley leaf and sheath waxes on agronomic performance in irrigated and dryland environments. - Crop Sci. 23: 116-120, 1983.

11271 - **BAGNALL, D., WOLFE, J., KING, R.W.:** Chill-induced wilting and hydraulic recovery in mung bean plants. - Plant Cell Environ. 6: 457-464, 1983.

*11272 - **BAILEY, C., LARSON, D.W.:** Water quality and pH effects on *Umbilicaria mammulata* (Ach.) Tuck. - Bryologist 85: 431-437, 1982.

*11273 - **BAILEY, W.G., DAVIES, J.A.:** The effect of uncertainty in aerodynamic resistance on evaporation estimates from the combination model. - Boundary-Layer Meteorol. 20: 187-199, 1981.

11274 - **BAKRADZE, N.G., BALLA, Yu.I.:** O kristallizatsii vnutrikletochnoĭ vody v tkanyakh rasteniĭ. [Crystallization of intracellular water in plant tissues.] - Biofizika 28: 119-121, 1983. [In R, ab: E.]

11275 - **BAKRADZE, N.G., MOISTSRAPISHVILI, K.M., KESHELASHVILI, L.V.:** O protsesse kristallizatsii vnutrikletochnoĭ vody v tkanyakh vinogradnoĭ lozy pri zamorazhivanii. [Cristallization of intracellular water in grape tissues at freezing.] - Biofizika 28: 888-890, 1983. [In R, ab: E.]

11276 - **BALDOCCHI, D.D., VERMA, S.B., ROSENBERG, N.J.:** Microclimate in the soybean canopy. - Agr. Meteorol. 28: 321-337, 1983.

11277 - **BALDOCCHI, D.D., VERMA, S.B., ROSENBERG, N.J., BLAD, B.L., GARAY, A., SPECHT, J.E.:** Leaf pubescence effects on the mass and energy exchange between soybean canopies and the atmosphere. - Agron. J. 75: 537-543, 1983.

11278 - **BALDOCCHI, D.D., VERMA, S.B., ROSENBERG, N.J., BLAD, B.L., GARAY, A., SPECHT, J.E.:** Influence of water stress on the diurnal exchange of mass and energy between the atmosphere and a soybean canopy. - Agron. J. 75: 543-548, 1983.

11279 - **BALEK, J., ČERMÁK, J., KUČERA, J., PRAX, A.:** A direct method for forest transpiration measurement. - J. Hydrol. 66: 123-131, 1983.

11280 - **BALESTRINI, S., VARTANIAN, N.:** Rhizogenic activity during water stress-induced senescence in *Brassica napus* var. *oleifera*. - Physiol. vég. 21: 269-277, 1983.

11281 - **BALINA, N.V.:** Ul'trastruktura yader pyl'tsy, zakalennykh i nezakalennykh protiv zasukhi rasteniĭ fasoli pod vliyaniem sukhoveya. [Influence of atmospheric drought on ultrastructure of pollen kernels in drought hardened and non-hardened bean plants.] - Sel'skokhoz. Biol. 1983(3): 48-52, 1983. [In R, ab: E.]

11282 - **BALLIF, J.-L., DUTIL, P.:** Lysimétrie en sol de craie non remanié. I. - Drainage, évaporation et rôle du couvert végétal. Résultats 1973-1980. - Agronomie 3: 857-866, 1983.

*11283 - **BANNIKOVA, I.A., IZMAĬLOVA, N.N., MAKSIMOVICH, S.V.:** Vodnyĭ balans i produktivnost' raznotravno-zlakovoĭ Karaganovoĭ stepi v Vostochnom Khangae (MNR). [Water relations and productivity of Karagan steppe in Eastern Khangae (Mongolia) with different grass species.] - Ėkologiya 1982(3): 6-12, 1982. [In R.]

11284 - **BANYAL, S., RAI, V.K.:** Reversal of osmotic stress effects by gibberellic acid in *Brassica campestris*. Recovery of hypocotyl growth, protein and RNA levels in the presence of GA. - Physiol. Plant. 59: 111-114, 1983.

11285 - **BARBOSA FILHO, M.P., FAGERIA, N.K., STONE, L.F.:** Manejo d'água e calagem em relação à produtividade e toxicidade de ferro em arroz. [Water management and liming in relation to grain yield and iron toxicity in rice.] - Pesq. agropec. Bras. 18: 903-910, 1983. [In Port, ab: E.]

*11286 - **BARCLAY, A.M., CRAWFORD, R.M.M.:** Temperature and anoxic injury in pea seedlings. - J. exp. Bot. 32: 943-949, 1981.

11287 - **BARFIELD, B.J., NORMAN, J.M.:** Potential for plant environment modification. - Agr. Water Manage. 7: 73-88, 1983. Also in: STONE, J.F., WILLIS, W.O. (ed.): Plant Production and Management Under Drought Conditions. Pp. 73-88. Elsevier Scientific Publishing Company, Amsterdam 1983.

11288 - **BARFIELD, B.J., PALMER, W.L., HAAN, C.T.:** Simulating the economics of supplemental irrigation for corn. - In: RAPER, C.D., Jr., KRAMER, P.J. (ed.): Crop Reactions to Water and Temperature Stresses in Humid, Temperate Climates. Pp. 151-166. Westview Press, Boulder 1983.

11289 - **BARLOW, E.W.R., DONOVAN, G.R., LEE, J.W.:** Water relations and composition of wheat ears grown in liquid culture: effect of carbon and nitrogen. - Aust. J. Plant Physiol. 10: 99-108, 1983.

11290 - **BARNES, D.K.:** Managing root systems for efficient water use: breeding plants for efficient water use. - In: TAYLOR, H.M., JORDAN, W.R., SINCLAIR, T.R. (ed.): Limitations to Efficient Water Use in Crop Production. Pp. 127-136. Amer. Soc. Agron., Inc., Crop Sci. Soc. Amer., Inc., Soil Sci. Soc. Amer., Inc., Madison 1983.

11291 - **BARRETT, D.R., FREAN, M.L., CRESSWELL, C.F.:** C_3 and C_4 photosynthetic and anatomical forms of *Alloteropsis semialata* (R. Br.) Hitchcock. I. Variability in photosynthetic characteristics, water utilization efficiency and leaf anatomy. - Ann. Bot. 51: 801-809, 1983.

*11292 - **BARRETT, J.E., NELL, T.A.:** Evapotranspiration and transpiration for four foliage plants. - HortScience 17: 918-919, 1982.

*11293 - **BARRETT, J.W.H., SKOGERBOE, G.V.:** Crop production functions and the allocation and use of irrigation water. - Agr. Water Manage. 3: 53-64, 1980.

*11294 - **BARROS, R.S., MAESTRI, M., MOREIRA, R.C.:** Sources of assimilates for expanding flower buds of coffee. - Turrialba 32: 371-377, 1982.

*11295 - **BASTIANPILLAI, V.A., STARK, C., FRANKE, G.:** The influence of gibberellin (GA_3), auxin (I-NAA), and cytokinin (kinetin,6-FAP) on the salt tolerance of wheat under greenhouse conditions. - Beitr. trop. Landwirt. Veterinärmed. 20: 365-370, 1982.

11296 - **BATCHELOR, C.H., ROBERTS, J.:** Evaporation from the irrigation water, foliage and panicles of paddy rice in north-east Sri Lanka. - Agr. Meteorol. 29: 11-26, 1983.

11297 - **BAUER, A., BLACK, A.L.:** Sprouting in intact mature spikes of hard red spring wheat. - Agron. J. 75: 1016-1022, 1983.

11298 - **BAUER, H., MARTHA, P., KIRCHNER-HEISS, B., MAIRHOFER, I.:** The CO_2 compensation point of C_3 plants - a re-examination II. Intraspecific variability. - Z. Pflanzenphysiol. 109: 143-154, 1983.

11299 - **BAUER, V.:** Závislost intenzity fotosyntézy listů mokřadních druhů vrb na základních meteorologických faktorech. [Dependence of willow photosynthetic rate on basic meteorological factors.] - In: Dny Rostlinné Fyziologie III. Pp. 165-166. ČSBS, Praha 1983. [In Czech.]

11300 - **BEARD, J.S.:** Ecological control of the vegetation of southwestern Australia: moisture versus nutrients. - In: KRUGER, F.J., MITCHELL, D.T., JARVIS, J.U.M. (ed.): Mediterranean-Type Ecosystems. The Role of Nutrients. Pp. 66-73. Springer-Verlag, Berlin - Heidelberg - New York 1983.

11301 - **BECKER, M., LÉVY, G.:** Influence d'un dessèchement du sol sur la nutrition minérale de jeunes plantes de résineux. - Ann. Sci. forest. 40: 325-336, 1983.

*11302 - **BECWAR, M.R., BURKE, M.J.:** Winter hardiness limitations and physiography of woody timberline flora. - In: LI, P.H., SAKAI, A. (ed.): Plant Cold Hardiness and Freezing Stress. Mechanisms and Crop Implications. Volume 2. Pp. 307-323. Academic Press, New York 1982.

*11303 - **BECWAR, M.R., RAJASHEKAR, C., HANSEN BRISTOW, K.J., BURKE, M.J.:** Deep undercooling of tissue water and winter hardiness limitations in timberline flora. - Plant Physiol. 68: 111-114, 1981.

11304 - **BECWAR, M.R., STANWOOD, P.C., LEONHARDT, K.W.:** Dehydration effects on freezing characteristics and survival in liquid nitrogen of desiccation--tolerant and desiccation-sensitive seeds. - J. Amer. Soc. hort. Sci. 108: 613-618, 1983.

*11305 - **BECWAR, M.R., STANWOOD, P.C., ROOS, E.E.:** Dehydration effects on imbibitional leakage from desiccation-sensitive seeds. - Plant Physiol. 69: 1132-1135, 1982.

11306 - **BECWAR, M.R., WALLNER, S.J., BUTLER, J.D.:** Effect of water stress on *in vitro* heat tolerance of turfgrass leaves. - HortScience 18: 93-95, 1983.

11307 - **BEEVER, R.E.:** Osmotic sensitivity of fungal variants resistant to dicarboximide fungicides. - Trans. Brit. mycol. Soc. 80: 327-331, 1983.

*11308 - **BEN-AMOTZ, A., SUSSMAN, I., AVRON, M.:** Glycerol production by *Dunaliella*. - Experientia 38: 49-52, 1982.

11309 - **BEN-HAYYIM, G., KOCHBA, J.:** Aspects of salt tolerance in a NaCl-selected stable cell line of *Citrus sinensis*. - Plant Physiol. 72: 685-690, 1983.

*11310 - **BEN-JAACOV, J., NATANSON, G., HAGILADI, A.:** The use of a wind-controlled overhead irrigation system to prevent damage by wind-borne salts. - J. Amer. Soc. hort. Sci. 105: 833-835, 1980.

*11311 - **BEN-JAACOV, J., POOLE, R.T., CONOVER, C.A.:** Effect of nutrition, soil water content, and duration of storage on quality of *Dieffenbachia maculata* (Lodd.) G. Don 'Rudolph Roehrs'. - HortScience 17: 611-612, 1982.

*11312 - **BENNETT, J.M., HAMMOND, L.C.:** Soil moisture and genotypic effects on the heat tolerance of peanut. - Soil Crop Sci. Soc. Florida Proc. 41: 97-100, 1982.

11313 - **BENNETT, J.M., HAMMOND, L.C.:** Grain yields of several corn hybrids in response to water stresses imposed during vegetative growth stages. - Soil Crop Sci. Soc. Florida Proc. 42: 107-111, 1983.

*11314 - **BENNETT, K.J., McPHERSON, H.G., WARRINGTON, I.J.:** Effect of pretreatment temperature on response of photosynthesis rate in maize to current temperature. - Aust. J. Plant Physiol. 9: 773-781, 1982.

11315 - **BEN-YEHOSHUA, S., SHAPIRO, B., CHEN, Z.E., LURIE, S.:** Mode of action of plastic film in extending life of lemon and bell pepper fruits by alleviation of water stress. - Plant Physiol. 73: 87-93, 1983.

*11316 - **BENZ, L.C., DOERING, E.J., REICHMAN, G.A.:** Water table and irrigation effects on alfalfa grown on sandy soils. - Can. agr. Eng. 24: 71-75, 1982.

*11317 - **BERGMANN, H., ROTH, D., AUGSTEN, H., SCHWARZ, K.:** Der Einfluss eines zeitlich begrenzten Wasserdefizits auf die Wasserausnutzung von Kulturpflanzen insbesondere Sommergerste. - Arch. Acker- Pflanzenbau Bodenk. 26: 217-225, 1982.

*11318 - **BERGTROM, G., SCHALLER, M., EICKMEIER, W.G.:** Ultrastructural and biochemical bases of resurrection in the drought-tolerant vascular plant, *Selaginella lepidophylla*. - J. Ultrastruct. Res. 78: 269-282, 1982.

11319 - **BERINGER, H., HAEDER, H.E., LINDHAUER, M.:** Water relationships and incorporation of ^{14}C assimilates in tubers of potato plants differing in potassium nutrition. - Plant Physiol. 73: 956-960, 1983.

11320 - **BERKOWITZ, G.A., CHEN, C., GIBBS, M.:** Stromal acidification mediates *in vivo* water stress inhibition of nonstomatal-controlled photosynthesis. - Plant Physiol. 72: 1123-1126, 1983.

11321 - **BERKOWITZ, G.A., GIBBS, M.:** Reduced osmotic potential effects on photosynthesis. Identification of stromal acidification as a mediating factor. - Plant Physiol. 71: 905-911, 1983.

11322 - **BERKOWITZ, G.A., GIBBS, M.:** Reduced osmotic potential inhibition of photosynthesis. Site-specific effects of osmotically induced stromal acidification. - Plant Physiol. 72: 1100-1109, 1983

*11323 - **BERLINER, M.D., MARTINDALE, P.T.:** Autoradiography of penetration of [^{14}C] mannitol into *Cosmarium botrytis* cells and protoplasts. - Plant Sci. Lett. 20: 371-377, 1981.

*11324 - **BERRY, J.A., RAISON, J.K.:** Responses of macrophytes to temperature. - In: LANGE, O.L., NOBEL, P.S., OSMOND, C.B., ZIEGLER, H. (ed.): Physiological Plant Ecology I. Pp. 277-338. Springer-Verlag, Berlin - Heidelberg - New York 1981.

*11325 - **BEUKES, D.J., WEBER, H.W.:** The effects of irrigation at different soil water levels on the water use characteristics of apple trees. - J. hort. Sci. 57: 383-391, 1982.

11326 - **BEWLEY, J.D., LARSEN, K.M., PAPP, J.E.T.:** Water-stress-induced changes in the pattern of protein synthesis in maize seedling mesocotyls: a comparison with the effects of heat shock. - J. exp. Bot. 34: 1126-1133, 1983.

11327 - **BEWLEY, J.D., OLIVER, M.J.:** Responses to a changing environment at the molecular level: Does desiccation modulate protein synthesis at the transcriptional or translational level in a tolerant plant? - In: RANDALL, D.D., BLEVINS, D.G., LARSON, R.L., RAPP, B.J., PALLARDY, S.G., NELSON, C.J., POLACCO, J.C., GEORGE, M.F. (ed.): Current Topics in Plant Biochemistry and Physiology. Volume 2. Pp. 145-164. University of Missouri, Columbia 1983.

11328 - **BEYL, C.A., MITCHELL, C.A.:** Alteration of growth, exudation rate, and endogenous hormone profiles in mechanically dwarfed sunflower. - J. Amer. Soc. hort. Sci. 108: 257-262, 1983.

11329 - **BHARAMBE, P.R., VARADE, S.B.:** Effect of water submergence periods on yield and biochemical changes of cotton plant. - Ind. J. agr. Res. 53: 179-181, 1983.

11330 - **BHASKARAN, S., SMITH, R.H., SCHERTZ, K.:** Sodium chloride tolerant callus of *Sorghum bicolor* (L.) Moench. - Z. Pflanzenphysiol. 112: 459-463, 1983.

11331 - **BHULLAR, S.S., JENNER, C.F.:** Responses to brief periods of elevated temperature in ears and grains of wheat. - Aust. J. Plant Physiol. 10: 549-560, 1983.

11332 - **BIELORAI, H., MANTELL, A., MORESHET, S.:** Water relations of cotton. - In: KOZLOWSKI, T.T. (ed.): Water Deficits and Plant Growth. Volume VII. Additional Woody Crop Plants. Pp. 49-87. Academic Press, New York - London - Paris - San Diego - San Francisco - São Paulo - Sydney - Tokyo - Toronto 1983.

11333 - **BIELORAI, H., SHALHEVET, J., LEVY, Y.:** The effect of high sodium irrigation water on soil salinity and yield of mature grapefruit orchard. - Irrig. Sci. 4: 255-266, 1983.

11334 - **BIL', K.Ya., LYUBIMOV, V.Yu., TRUSOV, M.F., GEDEMOV, T., ATAKHONOV, B.O.:** Uchastie trekh tipov avtotrofnykh tkaneǐ v sutochnoǐ dinamike assimilyatsii CO_2 u nekotorykh predstaviteleǐ sem. *Chenopodiaceae*. [The participation of three types of autotrophic tissues in diurnal dynamics of CO_2 assimilation in some *Chenopodiaceae* species.] - Bot. Zh. 68: 54-61, 1983. [In R.]

11335 - **BILLINGS, W.D., LUKEN, J.O., MORTENSEN, D.A., PETERSON, K.M.:** Increasing atmospheric carbon dioxide: possible effects on arctic tundra. - Oecologia 58: 286-289, 1983.

*11336 - **BISSING, D.R.:** Evolution of leaf architecture in the chaparral species *Fremontodendron californicum* ssp. *californicum* (*Sterculiaceae*). - Amer. J. Bot. 69: 957-972, 1982.

11337 - **BISSON, M.A., KIRST, G.O.:** Osmotic adaptations of charophyte algae in the Coorong, South Australia and other Australian lakes. - Hydrobiologia 105: 45-51, 1983.

11338 - **BÍZIK, J.:** Ovplyvňovanie úrody ozimnej viky a kukurice hnojivami a závlahou v osevnom postupe. [Influencing the yields of winter vetch and maize by means of fertilizers and irrigation in crop rotation.] - Rost. Výroba (Praha) 29: 839-845, 1983. [In Slov, ab: R, E.]

*11339 - **BLACKMAN, D., WARE, S.:** Soil moisture and the distribution of *Quercus prinus* and *Quercus rubra*. - Castanea 47: 360-367, 1982.

11340 - **BLACKMAN, P.G., DAVIES, W.J.:** The effects of cytokinins and ABA on stomatal behaviour of maize and *Commelina*. - J. exp. Bot. 34: 1619-1626, 1983.

*11341 - **BLACKSHAW, R.E., STOBBE, E.H., SHAYKEWICH, C.F., WOODBURY, W.:** Influence of soil temperature and soil moisture on green foxtail (*Setaria viridis*) establishment in wheat (*Triticum aestivum*). - Weed Sci. 29: 179-184, 1981.

*11342 - **BLAD, B.L.:** Atmospheric demand for water. - In: TEARE, I.D., PEET, M.M. (ed.): Crop-Water Relations. Pp. 1-44. John Wiley & Sons, New York - Chichester - Brisbane - Toronto - Singapore 1982.

11343 - **BLAKE, T.J.:** Transplanting shock in white spruce; effect of cold-storage and root pruning on water relations and stomatal conditioning. - Physiol. Plant. 57: 210-216, 1983.

11344 - **BLANCO-LÓPEZ, M.A., JIMÉNEZ-DÍAZ, R.M.:** Effect of irrigation on susceptibility of sunflower to *Macrophomina phaseoli*. - Plant Dis. 67: 1214-1217, 1983.

11345 - **BLEKHMAN, G.I.:** RNKaznaya aktivnost' i depolimerizatsiya RNK u nezasukhoustoǐchivogo sorta pshenitsy Diamant pri obezvozhivanii i regidratatsii. [RNAase activity and RNA depolymerization in the drought-sensitive wheat variety Diamant during dehydration and rehydration.] - Fiziol. Rast. 30: 511-517, 1983. [In R, ab: E.]

11346 - **BLOCK, F.R., LENZ, F., ONKELS, R.:** Einrichtungen zur Messung von Photosynthese, Respiration und Transpiration. - Gartenbauwissenschaft 48: 128-134, 1983.

11347 - **BLUM, A.:** Breeding programs for improving crop resistance to water stress. - In: RAPER, C.D., KRAMER, P.J. (ed.): Crop Reactions to Water and Temperature Stresses in Humid, Temperate Climates. Pp. 263-275. Westview Press, Boulder 1983.

*11348 - **BLUM, A., EBERCON, A.:** Cell membrane stability as a measure of drought and heat tolerance in wheat. - Crop Sci. 21: 43-47, 1981.

*11349 - **BLUM, A., GOZLAN, G., MAYER, J.:** The manifestation of dehydration avoidance in wheat breeding germplasm. - Crop Sci. 21: 495-499, 1981.

11350 - **BLUM, A., MAYER, J., GOZLAN, G.:** Associations between plant production and some physiological components of drought resistance in wheat. - Plant Cell Environ. 6: 219-225, 1983.

11351 - **BLUMWALD, E., MEHLHORN, R.J., PACKER, L.:** Studies of osmoregulation in salt adaptation of cyanobacteria with ESR spin-probe techniques. - Proc. nat. Acad. Sci. USA 80: 2599-2602, 1983.

11352 - **BLUMWALD, E., MEHLHORN, R.J., PACKER, L.:** Ionic osmoregulation during salt adaptation of the cyanobacterium *Synechococcus* 6311. - Plant Physiol. 73: 377-380, 1983.

*11353 - **BLUMWALD, E., TEL-OR, E.:** Osmoregulation and cell composition in salt--adaptation of *Nostoc muscorum*. - Arch. Microbiol. 132: 168-172, 1982.

11354 - **BOAG, S., PORTIS, A.R.:** Inhibition of the light activation of fructose-1,6--bisphosphatase and sedoheptulose-1,7-bisphosphatase by osmotic stress in isolated spinach chloroplasts. - In: Sixth International Congress on Photosynthesis. Abstracts. Volume 1. Pp. 74. Brussels 1983.

11355 - **BÖCHER, T.W.:** The allotetraploid *Saxifraga nathorsti* and its probable progenitors *S. aizoides* and *S. oppositifolia*. - Meddelelser Grønland, Bioscience 11: 3-22, 1983.

11356 - **BODDY, L., RAYNER, A.D.M.:** Origins of decay in living deciduous trees: the role of moisture content and a re-appraisal of the expanded concept of tree decay. - New Phytol. 94: 623-641, 1983.

11357 - **BODE, J., WILD, A.:** Influence of CCC on content and activity of RuBPC-ase, content of soluble reducing sugars and water relations in developing wheat leaves. - In: Sixth International Congress on Photosynthesis. Abstracts. Volume 2. Pp. 291. Brussels 1983.

11358 - **BOĬKOV, S.:** Nyakoi korelativni i regresionni zavisimosti pri napoyavana lyutserna. [Correlations and regressions in irrigated alfalfa.] - Rasteniev. Nauki 20(1): 66-76, 1983. [In Bulg, ab: R, E.]

11359 - **BOIS, J.F., COUCHAT, P.:** Comparison of the effects of water stress on the root systems of two cultivars of upland rice (*Oryza sativa* L.). - Ann. Bot. 52: 479-487, 1983.

11360 - **BOLGIANO, N.C., SAFIR, G.R., WARNCKE, D.D.:** Mycorrhizal infection and growth of onion in the field in relation to phosphorus and water availability. - J. Amer. Soc. hort. Sci. 108: 819-825, 1983.

11361 - **BONANNO, A.R., MACK, H.J.:** Use of canopy-air temperature differentials as a method for scheduling irrigations in snap beans. - J. Amer. Soc. hort. Sci. 108: 826-831, 1983.

11362 - **BONANNO, A.R., MACK, H.J.:** Yield components and pod quality of snap beans grown under differential irrigation. - J. Amer. Soc. hort. Sci. 108: 832-836, 1983.

11363 - **BONANNO, A.R., MACK, H.J.:** Water relations and growth of snap beans as influenced by differential irrigation. - J. Amer. Soc. hort. Sci. 108: 837-844, 1983.

*11364 - **BOOTE, K.J.**: Peanut. - In: TEARE, I.D., PEET, M.M. (ed.): Crop-Water Relations. Pp. 255-286. John Wiley & Sons, New York - Chichester - Brisbane - Toronto - Singapore 1982.

*11365 - **BOOTE, K.J., BENNETT, J.M.**: Teaching techniques in crop water relations, gas exchange, and growth analysis. - J. agron. Educ. 11: 13-16, 1982.

*11366 - **BORISOVA, I.V., ANISIMOVA, K.I., BESPALOVA, Z.G., BOBROVSKAYA, N.I., KAZANTSEVA, T.I., POPOVA, T.A., SLEMNEV, N.N.**: *Salsola passerina (Chenopodiaceae)* v severnoĭ Gobi (MNR). [*Salsola passerina (Chenopodiaceae)* in the northern Gobi (Mongolia).] - Bot. Zh. 67: 1196-1206, 1982. [In R, ab: E.]

*11367 - **BORISOVA, T.A., LAZAREVA, N.P., ZHOLKEVICH, V.N.**: Vliyanie khimicheskikh agentov na ėndodermal'nyĭ skachok vodnogo potentsiala i ėksudatsiu korneĭ *Zea mays* L. [Effect of chemical agents on the endodermal jump of water potential and the exudation in roots of *Zea mays* L.] - Dokl. Akad. Nauk SSSR 267: 766-768, 1982. [In R.]

*11368 - **BÖRNER, T., FÖRSTER, H.**: Zum Einfluss des Plastoms auf Plastidenzahl und Zellform bei der Sorte "Mrs. Parker" von *Pelargonium zonale* Hort. - Wiss. Z. M.-Luther-Univ. Halle-Wittenberg, Math.-naturwiss. Reihe 30 (3): 77-81, 1981.

11369 - **BORNMAN, C.H.**: Morphological, anatomical and ultrastructural responses to abscisic acid. - In: ADDICOTT, F.T.(ed.): Abscisic Acid. Pp. 523-551, Praeger Publishers, New York 1983.

11370 - **BOŠNJAK, D.**: Evaporacija sa slobodne vodene površine kao osnova zalivnog režima i njen odnos prema ETP kukuruza i soje. [Evaporation from open water surface as a basis of irrigation schedule and its relation to the ETP of corn and soybean.] - Arh. poljoprivredne Nauke 44: 323-344, 1983. [In Croat, ab: E.]

11371 - **BOSSELAERS, J.P.**: Cytokinin effects on leaf architecture in *Phaseolus vulgaris* L. - J. exp. Bot. 34: 1007-1017, 1983.

11372 - **BOT, G.P.A.**: Greenhouse Climate: From Physical Processes to a Dynamic Model. - Landbouwhogeschool, Wageningen 1983.

*11373 - **BOTHA, C.E.J., EVERT, R.F., CROSS, R.H.M., MARSHALL, D.M.**: The suberin lamella, a possible barrier to water movement from the veins to the mesophyll of *Themeda triandra* Forsk. - Protoplasma 112: 1-8, 1982.

11374 - **BOTTACIN, A., SACCOMANI, M., SPETTOLI, P., CACCO, G.**: NaCl-induced modifications of nitrogen absorption and assimilation in salt tolerant and salt resistant millet ecotypes (*Pennisetum typhoideum* L. Rich.). - In: SARIĆ, M.R., LOUGHMAN, B.C. (ed.): Genetic Aspects of Plant Nutrition. Pp. 203-207. Martinus Nijhoff Publishers, The Hague - Boston - Lancaster 1983.

11375 - **BOUILLARD, L., LE RUDULIER, D.**: Nitrogen fixation under osmotic stress: enhancement of nitrogenase biosynthesis in *Klebsiella pneumonie* by glycine betaine. - Physiol. vég. 21: 447-457, 1983.

*11376 - **BOUTHYETTE, P.-Y., JAGENDORF, A.T.**: Oligomycin effects on ATPase and photophosphorylation of pea chloroplast thylakoid membranes. - Plant Physiol. 69: 888-896, 1982.

*11377 - **BOYER, J.S.**: Plant productivity and environment. - Science 218: 443-448, 1982.

11378 - **BOYER, J.S.:** Molecular aspects of photosynthesis at low leaf water potentials. - In: Sixth International Congress on Photosynthesis. Abstracts. Volume 1. Pp. 17. Brussels 1983.

11379 - **BOYER, J.S.:** Subcellular mechanisms of plant response to low water potential. - Agr. Water Manage. 7: 239-248, 1983. Also in: STONE, J.F., WILLIS, W.O. (ed.): Plant Production and Management under Drought Conditions. Development in Agricultural and Managed-Forest Ecology. Vol. 12. Pp. 239-248. Elsevier Scientific Publishing Company, Amsterdam 1983.

11380 - **BOYER, J.S.:** Environmental stress and crop yields. - In: RAPER, C.D., KRAMER, P.J. (ed.): Crop Reactions to Water and Temperature Stresses in Humid, Temperate Climates. Pp. 3-7. Westview Press, Boulder 1983.

11381 - **BOYER, J.S., YOUNIS, H.M.:** Molecular aspects of photosynthesis at low leaf water potentials. - In: MARCELLE, R., CLIJSTERS, H., VAN POUCKE, M. (ed.): Effects of Stress on Photosynthesis. Pp. 29-33. Martinus Nijhoff / Dr W. Junk Publishers. The Hague - Boston - London 1983.

11382 - **BRADBURY, I.K., GRACE, J.:** Primary production in wetlands. - In: GORE, A.J.P. (ed.): Mires: Swamp, Bog, Fen and Moor. Part A. General Studies. Pp. 285-310. Elsevier Scientific Publishing Company, Amsterdam 1983.

11383 - **BRADFORD, K.J.:** Water relations and growth of the *flacca* tomato mutant in relation to abscisic acid. - Plant Physiol. 72: 251-255, 1983.

11384 - **BRADFORD, K.J.:** Effects of soil flooding on leaf gas exchange of tomato plants. - Plant Physiol. 73: 475-479, 1983.

11385 - **BRADFORD, K.J.:** Involvement of plant growth substances in the alteration of leaf gas exchange of flooded tomato plants. - Plant Physiol. 73: 480-483, 1983.

11386 - **BRADFORD, K.J., SHARKEY, T.D., FARQUHAR, G.D.:** Gas exchange, stomatal behavior, and $\delta^{13}C$ values of the *flacca* tomato mutant in relation to abscisic acid. - Plant Physiol. 72: 245-250, 1983.

11387 - **BRAUN, H.J.:** Zur Dynamik des Wassertransportes in Bäumen. - Ber. Deut. bot. Ges. 96: 29-47, 1983.

11388 - **BRAUNE, W., LEMAN, A., TAUBERT, H.:** Pflanzenanatomisches Praktikum I. - VEB Gustav Fischer Verlag, Jena 1983.

11389 - **BRENNINGER, C., TRANQUILLINI, W.:** Photosynthese, Transpiration und Spaltöffnungsverhalten verschiedener Holzarten nach Begasung mit SO_2. - Europ. J. Forest Pathol. 13: 228-238, 1983.

*11390 - **BRERETON, A.J., KEANE, T.:** The effect of water on grassland productivity in Ireland. - Irish J. agr. Res. 21: 227-248, 1982.

*11391 - **BRESSAN, R.A., HANDA, A.K., HANDA, S., HASEGAWA, P.M.:** Growth and water relations of cultured tomato cells after adjustment to low external water potentials. - Plant Physiol. 70: 1303-1309, 1982.

11392 - **BREŽNÝ, O.:** Zvláštnosti závlahového režimu skorých zemiakov pestovaných na ľahkých pôdach. [The peculiarities of the irrigation regime for early potatoes grown on light-textures soils.] - Rost. Výroba (Praha) 29: 787-793, 1983. [In Slov, ab: R, E.]

*11393 - **BRINGFELT, B.:** A forest evapotranspiration model using synoptic data. - Swedish Meteorological and Hydrological Institute Rep. Meteorol. Climatol. 36: 1-62, 1982.

*11394 - **BRINGFELT, B.:** Air humidity and radiation influence on forest transpiration. - Agr. Meteorol. 26: 297-307, 1982.

*11395 - **BRISTOW, K.L., De JAGER, J.M.:** Leaf water potential measurements using a strip chart recorder with the leaf psychrometer. - Agr. Meteorol. 22: 149-152, 1980.

11396 - **BROWN, A.D.:** Halophylic prokaryotes. - In: LANGE, O.L., NOBEL, P.S., OSMOND, C.B., ZIEGLER, H. (ed.): Physiological Plant Ecology III. Responses to the Chemical and Biological Environment. Pp. 137-162. Springer-Verlag, Berlin - Heidelberg - New York 1983.

*11397 - **BROWN, L.M.:** Photosynthetic and growth responses to salinity in a marine isolate of *Nannochloris bacillaris* (*Chlorophyceae*). - J. Phycol. 18: 483-488, 1982.

*11398 - **BROWN, P.H., OUTLAW, W.H., Jr.:** Effect of fusicoccin on dark $^{14}CO_2$ fixation by *Vicia faba* guard cell protoplasts. - Plant Physiol. 70: 1700-1703, 1982.

11399 - **BROWN, P.W., TANNER, C.B.:** Alfalfa osmotic potential: A comparison of the water-release curve and frozen-tissue methods. - Agron. J. 75: 91-93, 1983.

11400 - **BROWN, P.W., TANNER, C.B.:** Alfalfa stem and leaf growth during water stress. - Agron. J. 75: 799-805, 1983.

11401 - **BROWNLEE, C., DUDDRIDGE, J.A., MALIBARI, A., READ, D.J.:** The structure and function of mycelial systems of ectomycorrhizal roots with special reference to their role in forming inter-plant connections and providing pathways for assimilate and water transport. - Plant Soil 71: 433-443, 1983.

11402 - **BRUCKLER, L.:** Rôle des propriétés physiques du lit de semences sur l'imbibition et la germination. I. Elaboration d'un modèle du système "terre-graine". - Agronomie 3: 213-222, 1983.

11403 - **BRUCKLER, L.:** Rôle des propriétés physiques du lit de semences sur l'imbibition et la germination. II. Contrôle expérimental d'un modèle d'imbibition des semences et possibilités d'applications. - Agronomie 3: 223-232, 1983.

11404 - **BRUMME, B., PULS, K.E., EGGERS, H.:** Verdunstung im Gewächshaus. - Gartenbauwissenschaft 48: 182-187, 1983.

11405 - **BÜLOW-OLSEN, A.:** Germination response to salt in *Festuca rubra* in a population from a salt marsh. - Holarctic Ecol. 6: 194-198, 1983.

11406 - **BULYCHEV, A.A., NIYAZOVA, M.M., TUROVETSKIĬ, V.B.:** Vliyanie deĭterirovaniya na chustvitel'nost' élektronotransportnykh reaktsiĭ v khloroplastakh k deĭstviyu gramitsidina. [Effect of deuteration on the gramicidin-sensitive electron transport reactions in chloroplasts.] - Biokhimiya 48: 857-860, 1983. [In R, ab: E.]

11407 - **BUNCE, J.A.:** Differential sensitivity to humidity of daily photosynthesis in the field in C_3 and C_4 species. - Oecologia 57: 262-265, 1983.

11408 - **BURSCHKA, C., TENHUNEN, J.D., HARTUNG, W.:** Diurnal variations in abscisic acid content and stomatal response to applied abscisic acid in leaves of irrigated and non-irrigated *Arbutus unedo* plants under naturally fluctuating environmental conditions. - Oecologia 58: 128-131, 1983.

11409 - **BURTON, G.W.:** Breeding programs for stress tolerance in forage and pasture crops. - In: RAPER, C.D., Jr., KRAMER, P.J. (ed.): Crop Reactions to Water and Temperature Stresses in Humid, Temperate Climates. Pp. 289-296. Westview Press, Boulder 1983.

11410 - **BUSCHMANN, C., BUCHANAN-BOLLIG, I.C.:** Changes of the chlorophyll fluorescence induction kinetics of C_3 and CAM plants during day/night cycles. - Photosynthesis Res. 4: 337-349, 1983.

11411 - **BUSONI, E., KOWALIK, P., SANESI, G.:** Effect of drainage conditions on winter wheat production. - Agr. Water Manage. 7: 425-438, 1983.

11412 - **BUVAT, R., ROBERT, G.:** Cytologie ultrastructurale de l'hydratation germinative des primordiums de racines des semences de l'Orge (*Hordeum vulgare* L.). II. La réhydratation proprement dite. - Ann. Sci. Natur. Bot. (Paris) 5: 49-70, 1983.

11413 - **BYTNEROWICZ, A., TAYLOR, O.C.:** Influence of ozone, sulfur dioxide, and salinity on leaf injury, stomatal resistance, growth, and chemical composition of bean plants. - J. environ. Quality 12: 397-405, 1983.

11414 - **CALDWELL, M.M., DEAN, T.J., NOWAK, R.S., DZUREC, R.S., RICHARDS, J.H.:** Bunchgrass architecture, light interception, and water-use efficiency: assessment by fiber optic point quadrats and gas exchange. - Oecologia 59: 178-184, 1983.

11415 - **CAMPBELL, C.A., DAVIDSON, H.R., McCAIG, T.N.:** Disposition of nitrogen and soluble sugars in Manitou spring wheat as influenced by N fertilizer, temperature and duration and stage of moisture stress. - Can. J. Plant Sci. 63: 73-90, 1983.

11416 - **CAO, Y.-Z., LU, Z.-S.:** [Accumulation of natural growth inhibiting substances under stress conditions and its relation to adaptation of plants to unfavorable environments.] - Acta bot. Sinica 25: 123-130, 1983. [In Chin, ab: E.]

11417 - **CAPE, J.N.:** Contact angles of water droplets on needles of Scots pine (*Pinus sylvestris*) growing in polluted atmospheres. - New Phytol. 93: 293-299, 1983.

*11418 - **CAPORN, S.J.M., LUDWIG, L.J., FLOWERS, T.J.:** Potassium deficiency and photosynthesis in tomato. - In: SCAIFE, A. (ed.): Plant Nutrition. (Proc. Ninth Int. Plant Nutr. Colloq.). Pp. 78-83. Commonwealth Agricultural Bureaux, Farnham Royal 1982.

*11419 - **CARBON, B.A., BARTLE, G.A., MURRAY, A.M.:** Patterns of water stress and transpiration in jarrah (*Eucalyptus marginata* Don ex Sm.) forests. - Aust. Forest Res. 11: 191-200, 1981.

11420 - **CARBONNEAU, A.:** Measurement of gross photosynthesis under natural conditions by using a $^{14}CO_2$ diffusion porometer 1. Description of some convenient techniques. - Photosynthetica 17: 235-239, 1983.

11421 - **CARBONNEAU, A., SMART, R.E., De LOTH, C.:** Measurement of gross photosynthesis under natural conditions by using a $^{14}CO_2$ diffusion porometer 2. Combined influences of some major ecological factors and leaf area on grapevine photosynthesis. - Photosynthetica 17: 426-430, 1983.

11422 - **CARLSON, R.W.:** The effect of SO_2 on photosynthesis and leaf resistance at varying concentrations of CO_2. - Environ. Pollut. Ser. A. 30: 309-321, 1983.

11423 - **CARLSON, R.W.:** Interaction between SO_2 and NO_2 and their effects on photosynthetic properties of soybean *Glycine max*. - Environ. Pollut. Ser. A 32: 11-38, 1983.

11424 - **CARMI, A., HESKETH, J.D., ENOS, W.T., PETERS, D.B.:** Interrelationships between shoot and photosynthesis, as affected by root growth restriction. Photosynthetica 17: 240-245, 1983.

*11425 - **CARPENTER, J.R., MITCHELL, C.A.:** Root respiration characteristics of flood--tolerant and flood-intolerant tree species. - J. Amer. Soc. hort. Sci. 105: 684-687, 1980.

11426 - **CARR, M.K.V.:** The irrigation of vegetables: identifying the need. - Scientia Hort. 34: 54-65, 1983.

11427 - **CARR, M.K.V., DODDS, S.M.:** Some effects of soil compaction on root growth and water use of lettuce. - Exp. Agr. 19: 117-130, 1983.

11428 - **CARTER, P.R., SHEAFFER, C.C.:** Alfalfa response to soil water deficits. I. Growth, forage quality, yield, water use, and water-use efficiency. - Crop Sci. 23: 669-675, 1983.

11429 - **CARTER, P.R., SHEAFFER, C.C.:** Alfalfa response to soil water deficits. II. Plant water potential, leaf conductance, and canopy temperature relationships. - Crop Sci. 23: 676-680, 1983.

11430 - **CARTER, P.R., SHEAFFER, C.C.:** Alfalfa response to soil water deficits. III. Nodulation and N_2 fixation. - Crop Sci. 23: 985-990, 1983.

11431 - **CARVER, T.L.W., BUSHNELL, W.R.:** The probable role of primary germ tubes in water uptake before infection by *Erysiphe graminis*. - Physiol. Plant Pathol. 23: 229-240, 1983.

11432 - **CARY, J.W., FISHER, H.D.:** Irrigation decisions simplified with electronics and soil water sensors. - Soil Sci. Soc. Amer. J. 47: 1219-1223, 1983.

11433 - **CASSEL, D.K.:** Effects of soil water characteristics and tillage practices on water storage and its availability to plant roots. - In: RAPER, C.D., Jr., KRAMER, P.J. (ed.): Crop Reactions to Water and Temperature Stresses in Humid, Temperate Climates. Pp. 167-186. Westview Press, Boulder 1983.

11434 - **CASTLEBERRY, R.M.:** Breeding programs for stress tolerance in corn. - In: RAPER, C.D., Jr., KRAMER, P.J. (ed.): Crop Reactions to Water and Temperature Stresses in Humid, Temperate Climates. Pp. 277-287. Westview Press, Boulder 1983.

*11435 - **CATLIN, P.B., OLSSON, E.A., ROST, T.J.:** Response of jojoba seedlings to waterlogged root environments. - HortScience 17: 894-895, 1982.

11436 - **CAVALIÉ, G., COURTIADE, B., PIQUEMAL, M., MERRIEN, A., RANTY, B.:** Sunflower RuBP carboxylase: activity and content changes under different environmental conditions. - In: Sixth International Congress on Photosynthesis. Abstracts. Volume 2. Pp. 296. Brussels 1983.

11437 - **CAVALIERI, A.J.:** Proline and glycinebetaine accumulation by *Spartina alterniflora* Loisel. in response to NaCl and nitrogen in a controlled environment. - Oecologia 57: 20-24, 1983.

11438 - **ČERMÁK, J., KUČERA, J., SIMON, J., DUŠEK, V.:** The electric conductance of seedling stems and the water content of spruce and pine seedlings in the course of desiccation. - Biol. Plant. 25: 468-471, 1983.

11439 - **CEULEMANS, R., GABRIELS, R., IMPENS, I.:** Antitranspirant effects on transpiration, net CO_2 exchange rate and water-use efficiency of azalea. - Scientia Hort. 19: 125-131, 1983.

11440 - **CEULEMANS, R., GABRIËLS, R., IMPENS, I.:** Effect of fertilization level on some physiological, morphological and growth characteristics of *Ficus benjamina*. - Physiol. Plant. 59: 253-256, 1983.

11441 - **CEULEMANS, R., IMPENS, I.:** Nouvelles informations concernant l'absorption photosynthetique de CO_2 et la croissance de différentes plantes ornementales et horticoles. - Rev. Agr. 36: 1123-1143, 1983.

11442 - **CEULEMANS, R., IMPENS, I., NG, A.P.:** Photosynthesis, stomatal conductance and leaf water potential during water stress situations in young rubbertrees (*Hevea brasiliensis*) under tropical conditions. - In: MARCELLE, R., CLIJSTERS, H., VAN POUCKE, M. (ed.): Effects of Stress on Photosynthesis. Pp. 111-119. Martinus Nijhoff / Dr W. Junk Publishers, The Hague - Boston - London 1983.

11443 - **CHAFFEY, N.J.:** Epidermal structure in the ligule of rice (*Oryza sativa* L.). - Ann. Bot. 52: 13-21, 1983.

11444 - **CHALMERS, D.J., OLSSON, K.A., JONES, T.R.:** Water relations of peach trees and orchards. - In: KOZLOWSKI, T.T. (ed.): Water Deficits and Plant Growth. Volume VII. Additional Woody Crop Plants. Pp. 197-232. Academic Press, New York - London - Paris - San Diego - San Francisco - São Paulo - Sydney - Tokyo - Toronto 1983.

*11445 - **CHANG, H.-G., MARKAKIS, P.:** Effect of moisture content on aflatoxin production in barley. - Cereal Chem. 58: 89-91, 1981.

11446 - **CHARPENTIER, J.P., BONNET-MASIMBERT, M.:** Influence d'une réhydratation préalable sur la germination *in vitro* du pollen de douglas (*Pseudotsuga menziesii*) après conservation. - Ann. Sci. forest. 40: 309-317, 1983.

*11447 - **CHASE, A.R.:** Influence of irrigation method on severity of selected fungal leaf spots of foliage plants. - Plant Dis. 66: 673-674, 1982.

*11448 - **CHASSAGNE, M.-H., GAUDILLÈRE, J.-P., MONTIES, B.:** Effets du rayonnement ultraviolet sur les plantes cultivées sous éclairement naturel et artificiel. II. - Effets sur la capacité photosynthétique chez la tomate et le poivron. - Acta Oecol. - Oecol. Plant. 2: 327-338, 1981.

*11449 - **CHATURVEDI, G.S., AGGARWAL, P.K., SINHA, S.K.:** Growth and yield of determinate and indeterminate cowpeas in dryland agriculture. - J. agr. Sci. 94: 137-144, 1980.

*11450 - **CHAVAN, P.D., KARADGE, B.A.:** Influence of salinity on mineral nutrition of peanut (*Arachis hypogea* L.). - Plant Soil 54: 5-13, 1980.

*11451 - **CHEEMA, S.S., MALHOTRA, O.P., SINGH. J.:** Nutrient uptake, yield, relative leaf water, chlorophyll and protein contents of wheat, as influenced by cycocel under limited water conditions. - J. Res. Punjab agr. Univ. 18: 125-132, 1981.

11452 - **CHEN, C.-L., SUNG, J.-M.:** Effect of water stress on the reduction of nitrate and nitrite by soybean nodules. - Plant Physiol. 73: 1065-1066, 1983.

11453 - **CHEN, T.H.H., YAMAMOTO, S.D.K., GUSTA, L.V., SLINKARD, A.E.:** Imbibitional chilling injury during chickpea germination. - J. Amer. Soc. hort. Sci. 108: 944-948, 1983.

*11454 - **CHETAL, S., WAGLE, D.S., NAINAWATEE, H.S.:** Phosphatidylethanolamine in wheat and barley leaves under water stress. - Phytochemistry 21: 1432-1433, 1982.

11455 - **CHETAL, S., WAGLE, D.S., NAINAWATEE, H.S.:** Phospholipid changes in wheat and barley chloroplast under water stress. - Plant Sci. Lett. 29: 273-278, 1983.

11456 - **CHIEN, Y.C., KAO, K.N.:** Effects of osmolality, cytokinin, and organic acids on pollen callus formation in *Triticale* anthers. - Can. J. Bot. 61: 639-641, 1983.

11457 - **CHIKOV, V.I., IVANOVA, N.P., ISFANDIYAROV, N.I.:** Issledovanie vodoobmena list'ev rasteniĭ s pomoshchyu kamery davleniya i membranoaktivnykh veshchestv. [Studies of water exchange in plant leaves by means of the pressure chamber and membrane-active substances.] - Fiziol. Biokhim. kul't. Rast. 15: 466-470, 1983. [In R, ab: E.]

11458 - **CHOUDHURY, B.:** Simulating the effects of weather variables and soil water potential on a corn canopy temperature. - Agr. Meteorol. 29: 169-182, 1983.

*11459 - **CHRISTENSEN, W.N., KILLORN, R.J.:** Wheat and barley growth and nitrogen fertilizer utilization under sprinkler irrigation. - Agron. J. 73: 307-312, 1981.

*11460 - **CHRISTY, A.L., PORTER, C.A.:** Canopy photosynthesis and yield in soybean. - In: GOVINDJEE (ed.): Photosynthesis. Volume II. Pp. 499-511. Academic Press, New York - London - Paris - San Diego - San Francisco - São Paulo - Sydney - Tokyo - Toronto 1982.

*11461 - **CHROUST, L.:** Intercepce kapalných srážek v husté a řídké smrkové mlazině. [Interception of liquid precipitations in dense and space spruce young stand.] - Práce VÚLHM (Praha) 60: 63-83, 1982. [In Czech, ab: E, R.]

*11462 - **CHUMA, Y., UCHIDA, S., SHEMSANGA, K.H.H.:** Bulk physical and thermal properties of cereal grains as affected by moisture content. - J. Fac. Agr. Kyushu Univ. 26: 57-70, 1981.

11463 - **CIONCO, R.M.:** Canopy gas exchange: canopy flow modeling and its applications. - In: TAYLOR, H.M., JORDAN, W.R., SINCLAIR, T.R. (ed.): Limitations to Efficient Water Use in Crop Production. Pp. 203-213. Amer. Soc. Agron., Inc., Crop Sci. Soc. Amer., Inc., Soil Sci. Soc. Amer., Inc., Madison 1983.

11464 - **CLARKE, J.M.:** Differential excised-leaf water retention capabilities of *Triticum* cultivars grown in field and controlled environments. - Can. J. Plant Sci. 63: 539-541, 1983.

11465 - **CLEMENS, J., CAMPBELL, L.C., NURISJAH, S.:** Germination, growth and mineral ion concentrations of *Casuarina* species under saline conditions. - Aust. J. Bot. 31: 1-9, 1983.

*11466 - **CLOUGH, B.F., ANDREWS, T.J., COWAN, I.R.:** Physiological processes in mangroves. - In: CLOUGH, B.F. (ed.): Mangrove Ecosystems in Australia. Structure, Function and Management. Pp. 193-210. Australian National University Press, Canberra 1982.

11467 - **CLOUTIER, Y.:** Changes in the electrophoretic patterns of the soluble proteins of winter wheat and rye following cold acclimation and desiccation stress. - Plant Physiol. 71: 400-403, 1983.

11468 - **CLOUTIER, Y., SIMINOVITCH, D.:** L'induction de la résistance au froid par la sécheresse: une méthode rapide et economique pour tester et selectionner des cultivars résistant de céréales d'hiver. - Can. J. Plant Sci. 63: 573-580, 1983.

11469 - **COCHRANE, M.P.:** Morphology of the crease region in relation to assimilate uptake and water loss during caryopsis development in barley and wheat. - Aust. J. Plant Physiol. 10: 473-491, 1983.

11470 - **COCKBURN, W.:** Stomatal mechanism as the basis of the evolution of CAM and C_4 photosynthesis. - Plant Cell Environ. 6: 275-279, 1983.

11471 - **COGELS, O.G.:** An irrigation system uniformity function relating the effective uniformity of water application to the scale of influence of the plant root zones. - Irrig. Sci. 4: 289-299, 1983.

11472 - **COHEN, Y., FUCHS, M., COHEN, S.:** Resistance to water uptake in a mature citrus tree. - J. exp. Bot. 34: 451-460, 1983.

*11473 - **COLSON, M.:** Influence d'une déshydratation et d'une congélation (-196 °C) sur les ultrastructures des plantules d'*Allium cepa*. Étude après imprégnation métallique. - Can. J. Bot. 60: 1545-1553, 1982.

*11474 - **CONN, J.S., SNYDER-CONN, E.K.:** The relationship of the rock outcrop microhabitat to germination, water relations, and phenology of *Erythrina flabelliformis* (*Fabaceae*) in southern Arizona. - Southwest. Natur. 25: 443-451, 1981.

*11475 - **CONRADO, L.L., SHANAHAN, R., EINSINGER, W.:** Effects of pH, osmolarity, and oxygen on solution uptake by cut rose flowers. - J. Amer. Soc. hort. Sci. 105: 680-683, 1980.

11476 - **COOKE, J.R.:** Water transport and balance within the plant: stomatal mechanics and gas exchange. - In: TAYLOR, H.M., JORDAN, W.R., SINCLAIR, T.R. (ed.): Limitations to Efficient Water Use in Crop Production. Pp. 173-181. Amer. Soc. Agron., Inc., Crop Sci. Soc. Amer., Inc., Soil Sci. Soc. Amer., Inc., Madison 1983.

11477 - **COOPER, P.I., FULLER, R.J.:** A transient model of the interaction between crop, environment and greenhouse structure for predicting crop yield and energy consumption. - J. agr. Eng. Res. 28: 401-417, 1983.

11478 - **COOPER, P.J.M., KEATINGE, J.D.H., HUGHES, G.:** Crop evapotranspiration - a technique for calculation of its components by field measurements. - Field Crops Res. 7: 299-312, 1983.

11479 - **CORDONNIER, M.J., JOHNSTON, T.J.:** Effects of wastewater irrigation and plant and row spacing on soybean yield and development. - Agron. J. 75: 908-913, 1983.

11480 - **CORLEY, R.H.V.:** Photosynthesis and age of oil palm leaves. - Photosynthetica 17: 97-100, 1983.

11481 - **CORNIC, G.:** Non-stomatal inhibition of net CO_2 uptake by ABA. - In: Sixth International Congress on Photosynthesis. Abstracts. Volume 1. Pp. 59. Brussels 1983.

11482 - **CORNIC, G., MIGINIAC, E.:** Nonstomatal inhibition of net CO_2 uptake by ($\pm$) abscisic acid in *Pharbitis nil*. - Plant Physiol. 73: 529-533, 1983.

11483 - **CORNIC, G., PRIOUL, J.-L., LOUASON, G.:** Stomatal and non-stomatal contribution in the decline in leaf net CO_2 uptake during rapid water stress. - Physiol. Plant 58: 295-301, 1983.

11484 - **COSGROVE, D.J., CLELAND, R.E.:** Solutes in the free space of growing stem tissues. - Plant Physiol. 72: 326-331, 1983.

11485 - **COSGROVE, D.J., CLELAND, R.E.:** Osmotic properties of pea internodes in relation to growth and auxin action. - Plant Physiol. 72: 332-338, 1983.

11486 - **COTTIGNIES, A.:** Teneur en eau et dormance dans le bourgeon de Frêne. - Z. Pflanzenphysiol. 111: 133-139, 1983.

11487 - **COUPLAND, D.:** Influence of light, temperature and humidity on the translocation and activity of glyphosate in *Elymus repens* (= *Agropyron repens*). - Weed Res. 23: 347-355, 1983.

11488 - **COX, R.M.:** Sensitivity of forest plant reproduction to long range transported air pollutants: *in vitro* sensitivity of pollen to simulated acid rain. - New Phytol. 95: 269-276, 1983.

11489 - **COXSON, D.S., BROWN, D., KERSHAW, K.A.:** The interaction between CO_2 diffusion and the degree of thallus hydration in lichens: some further comments. - New Phytol. 93: 247-260, 1983.

11490 - **COXSON, D.S., KERSHAW, K.A.:** Rehydration response of nitrogenase activity and carbon fixation in terrestrial *Nostoc commune* from *Stipa-Bouteloa* grassland. - Can. J. Bot. 61: 2658-2668, 1983.

11491 - **CRAKER, L.E., SEIBERT, M., CLIFFORD, J.T.:** Growth and development of radish (*Raphanus sativus*, L.) under selected light environments. - Ann. Bot. 51: 59-64, 1983.

*11492 - **CRASSWELLER, R.M., FERREE, D.C.:** The effects of overtree misting for bloom delay on soil water status, net photosynthesis, transpiration, and carbohydrate levels of apple trees. - Ohio Agr. Res. Dev. Cent. Res. Circular 259: 11-15, 1980.

*11493 - **CREAMER, F.L., FOX, R.H.:** The toxicity of banded urea or diammonium phosphate to corn as influenced by soil temperature, moisture and pH. - Soil Sci. Soc. Amer. J. 44: 296-300, 1980.

11494 - **CRISTÓBAL, R., FI, J., PERZ, N.:** Diferencias morfológicas en hojas de tomate cultivado bajo dos niveles de radiación solar. [Morphological differences of leaves of tomato cultivated under two levels of solar radiation.] - Cienc. Agr. 16: 39-47, 1983. [In Span, ab: E.]

*11495 - **CRITCHLEY, C.:** Stimulation of photosynthetic electron transport in a salt--tolerant plant by high chloride concentration. - Nature 298: 483-485, 1982.

*11496 - **CRITCHLEY, C., BAIANU, I.C., GOVINDJEE, GUTOWSKY, H.S.:** The role of chloride in O_2 evolution by thylakoids from salt-tolerant higher plants. - Biochim. biophys. Acta 682: 436-445, 1982.

11497 - **CRUZ, R.T., O'TOOLE, J.C., MOODY, K.:** Leaf water potential of weeds and rice (*Oryza sativa*). - Weed Sci. 31: 410-414, 1983.

11498 - **CSONKA, L., Le RUDULIER, D., YANG, S.S., VALENTINE, A., CROUGHAN, T., STAVAREK, S.J., RAINS, D.W., VALENTINE, R.C.:** Genetic engineering for osmotically tolerant microorganisms and plants. - In: RAPER, C.D., Jr., KRAMER, P.J. (ed.): Crop Reactions to Water and Temperature Stresses in Humid, Temperate Climates. Pp. 245-261. Westview Press, Boulder 1983.

11499 - **CURE, J.D., RAPER, C.D., Jr., PATTERSON, R.P., JACKSON, W.A.:** Water stress recovery in soybeans as affected by photoperiod during seed development. - Crop Sci. 23: 110-115, 1983.

*11500 - **CURRAN, P.J.:** Multispectral remote sensing for estimating biomass and productivity. - In: SMITH, H. (ed.): Plants and the Daylight Spectrum. Pp. 65-99. Academic Press, London - New York 1981.

11501 - **CURRY, R.B., ESHEL, A.:** Use of simulation as a tool in crop management strategies for stress avoidance. - In: RAPER, C.D., Jr., KRAMER, P.J. (ed.): Crop Reactions to Water and Temperature Stresses in Humid, Temperate Climates. Pp. 231-242. Westview Press, Boulder 1983.

11502 - **DAESCHEL, M.A., FLEMING, H.P.:** Rapid and specific staining for routes of liquid entry into cucumber fruit. - J. Amer. Soc. hort. Sci. 108: 481-483, 1983.

11503 - **DAINES, R.J., MINOCHA, S.C.:** Regulation of phenylalanine ammonia-lyase in germinating lettuce seeds (*Lactuca sativa* L. cv. Grand Rapids): Effects of abscisic acid and water stress. - Z. Pflanzenphysiol. 110: 69-76, 1983.

11504 - **DALE, R.F.:** Temperature pertubations in the midwestern and southeastern United States important for corn production. - In: RAPER, C.D., Jr., KRAMER, P.J. (ed.): Crop Reactions to Water and Temperature Stresses in Humid, Temperate Climates. Pp. 21-32. Westview Press, Boulder 1983.

11505 - **DAMAŠKA, J., FOJTÍK, L.:** Matematický model proměnlivosti výnosu vyvolané hydrotermickými podmínkami. [A mathematical model of yield variability caused by weather conditions.] - Rost. Výroba (Praha) 29: 1307-1313, 1983. [In Czech, ab: R, E.]

*11506 - **DANNOWSKI, M., SCHÄFER, W., KÜNKEL, K.:** Standortspezifische Aussagen zur Energienutzung von Winterweizen im Freiland. - Arch. Acker- Pflanzenbau Bodenk. 25: 601-610, 1981.

11507 - **DANTUMA, G., THOMPSON, R.:** Whole-crop physiology and yield components. - In: HEBBLETHWAITE, P.D. (ed.): The Faba Bean (*Vicia faba* L.). A basis for improvement. Pp. 143-158. Butterworths, London - Boston - Durban - Singapore - Sydney - Toronto - Wellington 1983.

11508 - **DARNELL, R.L., FERREE, D.C.:** The influence of environment on apple tree growth, leaf wax formation, and foliar absorption. - J. Amer. Soc. hort. Sci. 108: 506-511, 1983.

*11509 - **DAS, V.S.R., RAJENDRUDU, G.:** The photosynthetic efficiency of flag leaf in relation to structural features in some crop plants. - Ind. J. Plant Physiol. 20: 123-128, 1977.

*11510 - **Da SILVA, P.R.F., STUTTE, C.A.:** Nitrogen volatilization from rice leaves. II. Effects of source of applied nitrogen in nutrient culture solution. - Crop Sci. 21: 913-916, 1981.

*11511 - **DAVENPORT, D.C., MARTIN, P.E., HAGAN, R.M.:** Evapotranspiration from riparian vegetation: Water relations and irrecoverable losses for saltcedar. - J. Soil Water Conserv. 37: 233-236, 1982.

*11512 - **DAVENPORT, D.C., MARTIN, P.E., HAGAN, R.M.:** Evapotranspiration from riparian vegetation: Conserving water by reducing saltcedar transpiration. - J. Soil Water Conserv. 37: 237-239, 1982.

*11513 - **DAVENPORT, D.C., URIU, K., HAGAN, R.M.:** Influence of crop load on the response of cherry fruit growth and size to antitranspirants. - HortScience 17: 217-218, 1982.

11514 - **DAVIES, F.S., ALBRIGO, L.G.:** Water relations of small fruits. - In: KOZLOWSKI, T.T. (ed.): Water Deficits and Plant Growth. Volume VII. Additional Woody Crop Plants. Pp. 89-136. Academic Press, New York - London - Paris - San Diego - San Francisco - São Paulo - Sydney - Tokyo - Toronto 1983.

11515 - **DAVIES, M.S., SINGH, A.K.:** Population differentiation in *Festuca rubra* L. and *Agrostis stolonifera* L. in response to soil waterlogging. - New Phytol. 94: 573-583, 1983.

11516 - **DAVIES, W.J., MANSFIELD, T.A.:** The role of abscisic acid in drought avoidance. - In: ADDICOTT, F.T. (ed.): Abscisic Acid. Pp. 237-268. Praeger Publishers, New York 1983.

11517 - **DAVIES, W.J., VAN VOLKENBURGH, E.:** The influence of water deficit on the factors controlling the daily pattern of growth of *Phaseolus* trifoliates. - J. exp. Bot. 34: 987-999, 1983.

*11518 - **DAWSON, R.F., SHRUM, J.E., Jr., MOHAMMED, R.A.:** Dependence of *Dioscorea* tuber growth rate on water content of an andept soil. - Soil Sci. Soc. Amer. J. 44: 1298-1302, 1980.

11519 - **DAY, W., LEGG, B.J.:** Water relations and irrigation response. - In: HEBBLETHWAITE, P.D. (ed.): The Faba Bean (*Vicia faba* L.). A Basis for Improvement. Pp. 217-231. Butterworths, London - Boston - Durban - Singapore - Sydney - Toronto - Wellington 1983.

11520 - **DE, R., BHEEMIAH, G., RAMSHESHIAH, K., RAO, Y.Y.:** Effect of mulches and antitranspirants on the grain yield of sorghum grown under limited irrigations on a deep vertisol. - J. agr. Sci. 100: 159-162, 1983.

11521 - **DE, R., RAO, Y.Y., IKRAMULLAH, M., RAO, L.G.G.:** Maize yield as affected by irrigation and evapotranspiration control treatments. - J. agr. Sci. 100: 731-734, 1983.

11522 - **DE, R., RAO, D.V.S.B., RAO, Y.Y., RAO, L.G.G., IKRAMULLAH, M.:** Modification of irrigation requirement of wheat through mulching and foliar application of transpiration suppressants. - Irrig. Sci. 4: 215-223, 1983.

*11523 - **DEAN, T.J., PALLARDY, S.G., COX, G.S.:** Photosynthetic responses of black walnut (*Juglans nigra*) to shading. - Can. J. Forest Res. 12: 725-730, 1982.

11524 - **DEANE-DRUMMOND, C.E., GLASS, A.D.M.:** Compensatory changes in ion fluxes into barley (*Hordeum vulgare* L. cv. Betzes) seedlings in response to differential root/shoot growth temperature. - J. exp. Bot. 34: 1711-1719, 1983.

11525 - **De ANGELIS, J., BERRY, R.E., KRANTZ, G.W.:** Photosynthesis, leaf conductance and leaf chlorophyll content in spider mite (*Acari: Tetranychidae*)-injured peppermint leaves. - Environ. Entomol. 12: 345-348, 1983.

11526 - **De ANGELIS, J.D., BERRY, R.E., KRANTZ, G.W.:** Evidence for spider mite (*Acari: Tetranychidae*) injury-induced leaf water deficits and osmotic adjustment in peppermint. - Environ. Entomol. 12: 336-339, 1983.

*11527 - **De ANGELIS, J.D., LARSON, K.C., BERRY, R.E., KRANTZ, G.W.:** Effects of spider mite injury on transpiration and leaf water status in peppermint. - Environ. Entomol. 11: 975-978, 1982.

11528 - **DeBUSK, T.A., RYTHER, J.H., WILLIAMS, L.D.:** Evapotranspiration of *Eichhornia crassipes* (Mart.) Solms and *Lemna minor* L. in central Florida: relation to canopy structure and season. - Aquatic Bot. 16: 31-39, 1983.

11529 - **DECKER, W.L.:** Probability of drought for humid and subhumid regions. - In: RAPER, C.D., Jr., KRAMER, P.J. (ed.): Crop Reactions to Water and Temperature Stresses in Humid, Temperate Climates. Pp. 11-19. Westview Press, Boulder 1983.

11530 - **De JONG, T.M.:** CO_2 assimilation characteristics of five *Prunus* tree species. - J. Amer. Soc. hort. Sci. 108: 303-307, 1983.

11531 - **DEMMIG, B., WINTER, K.:** Photosynthetic characteristics of chloroplasts isolated from *Mesembryanthemum crystallinum* L., a halophytic plant capable of Crassulacean acid metabolism. - Planta 159: 66-76, 1983.

*11532 - **De OLIVEIRA, M.O., Da MOTA, F.S., Da SILVA, J.B.:** Estimates of potential evapotranspiration (Penman) as a function of geographical factors in Brazil. - Agr. Meteorol. 22: 207-215, 1980.

*11533 - **De OLIVEIRA, S.L., MACEDO, M.M.C., PORTO, M.C.M.:** Efeito do déficit de água na produção de raízes de mandioca. [Effects of water stress on cassava root production.]- Pesq. agropec. Bras. 17: 121-124, 1982. [In Port, ab: E.]

11534 - **DERCO, M.:** Možnosti zmiernenia negatívneho účinku monokultúry ozimnej pšenice na úrodu zrna závlahou a ďalšími opatreniami. [The possibilities of reducing the adverse effect of winter wheat continuous growing on grain yield through irrigation and other practices.] - Rost. Výroba (Praha) 29: 805-816, 1983. [In Slov, ab: E, R.]

11535 - **DERCO, M., BARTA, V.:** Možnosti kompenzácie zníženia úrody cukrovej repy po nekompletnom vzídení rastlín v závislosti od vodného režimu pôdy. [The compensation of a decrease in sugar beet yield resulting from incomplete emergence of plants related to water regime.] - Rost. Výroba (Praha) 29: 27-35, 1983. [In Slov, ab: E, R.]

11536 - **DERCO, M., BARTA, V.:** Účinok organických hnojív na úrodu ozimnej pšenice pestovanej v monokultúre v závislosti od vodného režimu pôdy. [The effect of organic fertilizers on the yield of continually grown winter wheat in relation to soil water regime.] - Rost. Výroba (Praha) 29: 245-254, 1983. [In Slov, ab: E, R.]

11537 - **DERCO, M., RAKOVAN, J.:** Interakčný účinok závlahy a plytkej orby na úrodu poľných plodín v závislosti od úrovne hnojenia. [The interaction of irrigation and shallow poughing as influencing the yield of field crops in dependence on the level of fertilization.] - Rost. Výroba (Praha) 29: 71-78, 1983. [In Slov, ab: E, R.]

11538 - **DERYUGINA, T.F., NESTEROVICH, N.D.:** O morfologo-anatomicheskikh osobennostyakh khvoi (cheshuek) nekotorykh vidov roda kiparisovik (*Chamaecyparis* Spach.). [Morphological and anatomical features of the conifer needles (squama) of some *Chamaecyparis* Spach. species.] - Dokl. Akad. Nauk BSSR 27: 371-374, 1983. [In R, ab: E.]

*11539 - **DESHPANDE, R., NIMBALKAR, J.D.:** Effect of salt stress on the $^{14}CO_2$ fixation efficiency of isolated mesophyll cells. - Geobios 7: 91-92, 1980.

11540 - **DeTAR, W.R., KIBLER, D.F., GRENOBLE, D.W., DANIELS, R., COLE, R.H., TUKEY, L.D., HAMPSON, S.H., GELLER, S.C.:** Trickle irrigation vs. no-irrigation of five horticultural crops in Pennsylvania. - Trans. ASAE 26: 82-86, 1983.

*11541 - **DEVENDER REDDY, M., KRISHNAMURTHY, I., ANAND REDDY, K., VENKATACHARI, A.:** Evapotranspiration relationship with pan evaporation and evapotranspiration ratio of corn under different nitrogen levels and moisture regimes. - Agr. Water Manage. 3: 227-231, 1980/81.

11542 - **DEVITT, D., JURY, W.A., STERNBERG, P., STOLZY, L.H.:** Comparison of methods used to estimate evapotranspiration for leaching control. - Irrig. Sci. 4: 59-69, 1983.

11543 - **DIALLO, N., QUEIROZ-CLARET, C.:** Effet du chlorure de sodium sur le métabolisme carboné de *Cakile maritima* et *Kalanchoe blossfeldiana*. - Physiol. vég. 21: 621-632, 1983.

11544 - **DIAZ, R.A., MATTHIAS, A.D., HANKS, R.J.:** Evapotranspiration and yield estimation of spring wheat from canopy temperature. - Agron. J. 75: 805-810, 1983.

11545 - **DIETZ, K.-J., HEBER, U.:** Carbon dioxide gas exchange and the energy status of leaves of *Primula palinuri* under water stress. - Planta 158: 349-356, 1983.

*11546 - **DITTRICH, P., FISCHER, K.:** Accumulation and conversion of 3-O-methyl-D-glucose in guard cells of *Commelina communis*. - Z. Naturforsch. Sect. C (Biosci.) 35: 832-834, 1980.

*11547 - **DIVATE, M.R., PANDEY, R.M.:** Salt tolerance in grapes. III. Effects of salinity on chlorophyll, photosynthesis and respiration. - Ind. J. Plant Physiol. 24: 74-79, 1981.

*11548 - **DIXON, R.K., COX, G.S.:** Water deficits and root growth of ectomycorrhizal white oak seedlings. - Trans. Missouri Acad. Sci. 14: 178, 1980.

11549 - **DIXON, R.K., PALLARDY, S.G., GARRETT, H.E., COX, G.S., SANDER, I.L.:** Comparative water relations of container-grown and bare-root ectomycorrhizal and nonmycorrhizal *Quercus velutina* seedlings. - Can. J. Bot. 61: 1559-1565, 1983.

*11550 - **DOLEY, D.:** Tropical and subtropical forests and woodlands. - In: KOZLOWSKI, T.T. (ed.): Water Deficits and Plant Growth. Volume VI. Woody Plant Communities. Pp. 210-305. Academic Press, New York - London - Toronto - Sydney - San Francisco 1981.

11551 - **DOMINY, P.J., BAKER, N.R., MEEK, I.T.:** Time-dependent changes in the *in vitro* effects of sodium chloride on photosynthetic electron transport of isolated chloroplasts. - Physiol. Plant. 57: 163-168, 1983.

11552 - **DONOVAN, T.J., MEEK, B.D.:** Alfalfa responses to irrigation treatment and environment. - Agron. J. 75: 461-464, 1983.

*11553 - **DONYUSHKINA, E.A., FAL'KOVA, T.V.:** Temperaturnyĭ i vodnyĭ rezhim klematisa v usloviyakh yuzhnogo berega Kryma. [Temperature and water regime of clematis under conditions of southern coast of the Crimea.] - Byull. gosudarstv. Nikitskogo bot. Sada 1981(3): 84-87, 1981. [In R, ab: E.]

11554 - **DOWNTON, W.J.S.:** Osmotic adjustment during water stress protects the photosynthetic apparatus against photoinhibition. - Plant Sci. Lett. 30: 137-143, 1983.

*11555 - **DOWNTON, W.J.S., BJÖRKMAN, O., PIKE, C.S.:** Consequences of increased atmospheric concentrations of carbon dioxide for growth and photosynthesis of higher plants. - In: PEARMAN, G.I. (ed.): Carbon Dioxide and Climate: Australian Research. Pp. 143-151. Australian Academy of Science, Canberra 1980.

11556 - **DOWNTON, W.J.S., MILLHOUSE, J.:** Turgor maintenance during salt stress prevents loss of variable fluorescence in grapevine leaves. - Plant Sci. Lett. 31: 1-7, 1983.

11557 - **DOWNTON, W.J.S., MILLHOUSE, J.:** Osmotic adjustment and chlorophyll fluorescence in water- and salt-stressed plants. - In: Sixth International Congress on Photosynthesis. Abstracts. Volume 1. Pp. 381. Brussels 1983.

11558 - **DREIER, W.:** La teneur en proline et la résistance des plantes aux sels. Biol. Plant. 25: 81-87, 1983.

11559 - **DREIER, W.:** Vérification et utilisation d'un test élaboré de préselection des plantes tolérantes aux sels, à partir de la teneur en proline. - Biol. Plant. 25: 88-94, 1983.

11560 - **DREYER, E., DAUDET, F.A.:** Photosynthesis and transpiration of whole walnut trees: diurnal variations and effects of a severe drought. - In: Sixth International Congress on Photosynthesis. Abstracts. Volume 1. Pp. 382. Brussels 1983.

11561 - **DRYDEN, P., Van ALFEN, N.K.:** Use of the pressure bomb for hydraulic conductance studies. - J. exp. Bot. 34: 523-528, 1983.

11562 - **Du CLOUX, H.C., RICHAUD, C., VIVOLI, J.:** Effect of increasing atmospheric carbon dioxide on growth, photosynthesis, photorespiration and water use efficiency. - In: Sixth International Congress on Photosynthesis. Abstracts. Volume 2. Pp. 369. Brussels 1983.

11563 - **DUDA, U., KACPERSKA, A.:** Comparative studies on desiccation resistance in leaves of *Dactylis glomerata* L. and *Festuca arundinacea* Schreb. - Biol. Plant. 25: 124-128, 1983.

11564 - **DUDECK, A.E., SINGH, S., GIORDANO, C.E., NELL, T.A., McCONELL, D.B.:** Effect of sodium chloride on *Cynodon* turfgrasses. - Agron. J. 75: 927-930, 1983.

11565 - **DUGAS, W.A., ARKIN, G.F., JACKSON, B.S.:** Factors affecting simulated crop yield spatial extrapolation. - Trans. ASAE 26: 1440-1443, 1445, 1983.

*11566 - **DUKE, S.H., KAKEFUDA, G.:** Role of the testa in preventing cellular rupture during imbibition of legume seeds. - Plant Physiol. 67: 449-456, 1981.

11567 - **DUKE, S.O., WILLIAMS, R.D., MARKHART, A.H., III:** Interaction of moisture stress and three phenolic compounds on lettuce seed germination. - Ann. Bot. 52: 923-926, 1983.

*11568 - **DUNCAN, P.J., BILDERBACK, T.E.:** Effects of irrigation systems, gibberellic acid, and photoperiod on seed germination of *Kalmia latifolia* L. and *Rhododendron maximum* L. - HortScience 17: 916-917, 1982.

11569 - **DURLEY, R.C., KANNANGARA, T., SEETHARAMA, N., SIMPSON, G.M.:** Drought resistance of *Sorghum bicolor*. 5. Genotypic differences in the concentrations of free and conjugated abscisic, phaseic and indole-3-acetic acids in leaves of field-grown drought-stressed plants. - Can. J. Plant Sci. 63: 131-145, 1983.

11570 - **DURRANT, M.J., PAYNE, P.A., McLAREN, J.S.:** The use of water and some inorganic salt solutions to advance sugar beet seed. I. Laboratory studies. - Ann. appl. Biol. 103: 507-515, 1983.

11571 - **DURRANT, M.J., PAYNE, P.A., McLAREN, J.S.:** The use of water and some inorganic salt solutions to advance sugar beet seed. II. Experiments under controlled and field conditions. - Ann. appl. Biol. 103: 517-526, 1983.

11572 - **DUYNISVELD, W.H.M., RENGER, M., STREBEL, O.:** Vergleich von zwei Simulationsmodellen zur Ermittlung der Wasserhaushaltskomponenten in der ungesättigten Bodenzone. - Z. Deut. geol. Ges. 134: 679-686, 1983.

11573 - **DWELLE, R.B., HURLEY, P.J., PAVEK, J.J.:** Photosynthesis and stomatal conductance of potato clones (*Solanum tuberosum* L.). Comparative differences in diurnal patterns, response to light levels, and assimilation through upper and lower leaf surfaces. - Plant Physiol. 72: 172-176, 1983.

11574 - **EAMUS, D., FENTON, R., WILSON, J.M.:** Stomatal behaviour and water relations of chilled *Phaseolus vulgaris* L. and *Pisum sativum* L. - J. exp. Bot. 34: 434-441, 1983.

11575 - **EAMUS, D., WILSON, J.M.:** ABA levels and effects in chilled and hardened *Phaseolus vulgaris*. - J. exp. Bot. 34: 1000-1006, 1983.

11576 - **EASTIN, J.D., CASTLEBERRY, R.M., GERIK, T.J., HULTQUIST, J.H., MAHALAKSHMI, V., OGUNLELA, V.B., RICE, J.R.:** Physiological aspects of high temperature and water stress. - In: RAPER, C.D., Jr., KRAMER, P.J. (ed.): Crop Reactions to Water and Temperature Stresses in Humid, Temperate Climates. Pp. 91-112. Westview Press, Boulder 1983.

11577 - **EDWARDS, W.R.N., JARVIS, P.G.:** A method for measuring radial differences in water content of intact tree stems by attenuation of gamma radiation. - Plant Cell Environ. 6: 255-260, 1983.

*11578 - **EGLEY, G.H., PAUL, R.N., Jr.:** Development, structure and function of subpalisade cells in water impermeable *Sida spinosa* seeds. - Amer. J. Bot. 69: 1402-1409, 1982.

11579 - **EGLI, D.B., MECKEL, L., PHILLIPS, R.E., RADCLIFFE, D., LEGGETT, J.E.:** Moisture stress and N redistribution in soybean. - Agron. J. 75: 1027-1031, 1983.

*11580 - **EGUNYOMI, A.:** On the stomata of some tropical African mosses. - Lindbergia 8: 121-124, 1982.

11581 - **EHLERINGER, J.:** Ecophysiology of *Amaranthus palmeri*, a Sonoran desert summer annual. - Oecologia 57: 107-112, 1983.

11582 - **EHLERINGER, J.:** Characterization of a glabrate *Encelia farinosa* mutant: morphology, ecophysiology, and field observations. - Oecologia 57: 303-310, 1983.

11583 - **EHLERINGER, J., MOONEY, H.A.:** Productivity of desert and mediterranean-climate plants. - In: LANGE, O.L., NOBEL, P.S., OSMOND, C.B., ZIEGLER, H. (ed.): Physiological Plant Ecology IV. Ecosystem Processes: Mineral Cycling, Productivity and Mans' Influence. Pp. 205-231. Springer-Verlag, Berlin - Heidelberg - New York 1983.

11584 - **EHRLER, W.L.:** The transpiration ratios of *Agave americana* L.and *Zea mays* L. as affected by soil water potential. - J. Arid Environ. 6: 107-113, 1983.

11585 - **EIBACH, R., ALLEWELDT, G.:** Einfluss der Wasserversorgung auf Wachstum, Gaswechsel und Substanzproduktion traubentragender Reben. I. Das vegetative Wachstum. - Vitis 22: 231-240, 1983.

11586 - **EICKMEIER, W.G.:** Photosynthetic recovery of the resurrection plant *Selaginella lepidophylla* (Hook. and Grev.) Spring: effects of prior desiccation rate and mechanisms of desiccation damage. - Oecologia 58: 115-120, 1983.

11587 - **EISSENSTAT, D.M., MITCHELL, J.E.:** Effects of seeding grass and clover on growth and water potential of Douglas-fir seedlings. - Forest Sci. 29: 166-179, 1983.

11588 - **ELIÁŠ, P.:** Water relations pattern of understorey species influenced by sunflecks. - Biol. Plant. 25: 68-74, 1983.

*11589 - **EL KAROURI, M.O.H., MANSI, M.G.:** Performance of sorghum (*Sorghum vulgare*) and maize (*Zea mays*) as forages in irrigated saline soils of the Sudan. - Exp. Agr. 16: 431-436, 1980.

11590 - **ELKINAWY, M.:** Free and bound abscisic acid in relation to water potential in aging cotyledons of *Lupinus albus*. - Physiol. Plant. 57: 514-520, 1983.

*11591 - **EL-KOMOS, F., OSWAL, M.C., KHANNA, S.S.:** Response of pearl-millet to soil moisture content, soil moisture stress and soil moisture transmission coefficients. - Agrochimica 25: 452-458, 1982.

11592 - **ELLENSON, J.L., RABA, R.M.:** Gas exchange and phytoluminography of single red kidney bean leaves during periods of induced stomatal oscillations. A demonstration of an integrated, spatially resolving physiometric technique. - Plant Physiol. 72: 90-95, 1983.

11593 - **ELSNER, B., von:** A computer simulation of the crop microclimate in a greenhouse. - Gartenbauwissenschaft 48: 275-279, 1983.

11594 - **ELSTNER, E.F., KELLER, G., PARADIES, I.:** Contrasting effects of the cotton defoliant Thidiazuron and aminoethoxyvinylglycine, an inhibitor of ethylene formation, on stomatal aperture and on ethylene formation in bean (*Phaseolus vulgaris*) leaves. - Ber. Deut. bot. Ges. 96: 459-467, 1983.

*11595 - **EL-TAMZINI, M.I., SHALADAN, M.S., NIAZI, Z.M.:** Antitranspirant increases alsol effectiveness as chemical aid for harvesting olive. - HortScience 17: 965-966, 1982.

11596 - **EMBERGER, G., WELTY, R.E.:** Effect of soil water matric potential on resistance to *Fusarium oxysporum* f.sp. *medicaginis* in alfalfa. - Phytopathology 73: 208-212, 1983.

*11597 - **ENGLISH, M.J., NAKAMURA, B.C.:** Deficit irrigation - effects of irrigation frequency on crop yields. - In: Water, Energy and Economic Alternatives. Pp. 162-172. Irrigation Association, Silver Spring 1982.

11598 - **ENOCH, S., GLINKA, Z.:** Turgor-dependent membrane permeability in relation to calcium level. - Physiol. Plant. 59: 203-207, 1983.

11599 - **EPIPHANIO, J.C.N.:** Avalição da condição hídrica da cultura do milho através de sensoriamento remoto termal. [Evaluation of corn water condition: a thermal infrared remote sensing approach.] - Pesq. agropec. Bras. 18: 1233-1241, 1983. [In Port, ab: E.]

11600 - **EPSTEIN, E.:** Crops tolerant of salinity and other mineral stresses. - In: NUGENT, J., O'CONNOR, M. (ed.): Better Crops for Food (Ciba Foundation Symposium 97). Pp. 61-82. Pitman, London 1983.

11601 - **ERDMANN, N.:** Organic osmoregulatory solutes in blue-green algae. - Z. Pflanzenphysiol. 110: 147-155, 1983.

11602 - **ERREDE, L.A.:** Plant influence on water flow through artificial membranes. - Ann. Bot. 52: 23-29, 1983.

11603 - **ERREDE, L.A.:** Correlation of water uptake and root exudation. - Ann. Bot. 52: 373-380, 1983.

11604 - **ERREDE, L.A.:** Studies of water uptake from multiple sources. - Ann. Bot. 52: 839-847, 1983.

*11605 - **ESCHBACH, J.M., MASSIMINO, D., MENDOZA, A.M.R.:** Effet d'une carence en chlore sur la germination, la croissance et la photosynthèse du cocotier. - Oléagineux 37: 115-125, 1982.

*11606 - **ESPINOZA, W., Da SILVA, E.M., De SOUZA, O.C.:** Irrigação de trigo em solo de cerrado. [Response of wheat to irrigation in a "cerrado" soil.] - Pesq. agropec. Bras. 15: 107-115, 1980. [In Port, ab: E.]

*11607 - **ETCHEVERS, J.D., MORAGHAN, J.T., JENSEN, R.E.:** Water status of wheat tillers under stress conditions. - Turrialba 32: 43-50, 1982.

*11608 - **ETEJERE, E.O.:** Effects of EPTC, fluorodifen and monuron on transpiration and photosynthetic oxygen output in *Eupatorium odoratum* plants. - Weed Res. 22: 313-318, 1982.

*11609 - **EVANS, L.S.:** Biological effects of acidity in precipitation on vegetation: a review. - Environ. exp. Bot. 22: 155-169, 1982.

*11610 - **EVEN-CHEN, Z., WEINBAUM, S.A., PEARCY, R.W.:** High temperature effects on leaf resistance, leaf water potential and photosynthesis of non-bearing prune trees. - J. Amer. Soc. hort. Sci. 106: 216-219, 1981.

11611 - **EVERT, D.R., MULLINIX, B.G., Jr.:** Xylem water potential of peach trees infected with phony disease. - HortScience 18: 719-721, 1983.

11612 - **EZE, J.M.O., DUMBROFF, E.B., THOMPSON, J.E.:** Effects of temperature and moisture stress on the accumulation of abscisic acid in bean. - Physiol. Plant. 58: 179-183, 1983.

*11613 - **FAENSEN-THIEBES, A., OVERDIECK, D.:** Wirkungen des Cadmiums auf Veränderungen von CO_2-Gaswechsel und Transpiration bei *Phaseolus vulgaris* L. nach Ozon--Begasung. - Verhandl. Ges. Ökol. 9: 277-281, 1981.

11614 - **FARQUHAR, G.D., WETSELAAR, R., WEIR, B.:** Gaseous nitrogen losses from plants. - In: FRENEY, J.R., SIMPSON, J.R. (ed.): Gaseous Loss of Nitrogen from Plant-Soil Systems. Pp. 159-180. Martinus Nijhoff / Dr W. Junk Publishers, The Hague - Boston - Lancaster 1983

*11615 - **FAZYLOVA, S.F.:** O fotosinteticheskoĭ produktivnosti *Ceratoides ewersmanniana* (Stschegl. ex Losinsk.) Botsch. et Ikonn. (teresken), vyrashchennogo v usloviyakh razlichnogo vodnogo rezhima pochvy. [Photosynthetic productivity of *Ceratoides ewersmanniana* grown under various soil water relations.] - Uzb. biol. Zh. 1980(5): 21-24, 1980. [In R.]

11616 - **FELDHAKE, C.M., DANIELSON, R.E., BUTLER, J.D.:** Turfgrass evapotranspiration. I. Factors influencing rate in urban environments. - Agron. J. 75: 824-830, 1983.

11617 - **FELKER, P., CANNELL, G.H., CLARK, P.R., OSBORN, J.F., NASH, P.:** Biomass production of *Prosopis* species (mesquite), *Leucaena* and other leguminous trees grown under heat/drought stress. - Forest Sci. 29: 592-606, 1983.

11618 - **FELKER, P., CANNELL, G.H., OSBORN, J.F., CLARK, P.R., NASH, P.:** Effects of irrigation on biomass production of 32 *Prosopis* (mesquite) accessions. - Exp. Agr. 19: 187-198, 1983.

*11619 - **FELKER, P., CLARK, P.R., LAAG, A.E., PRATT, P.F.:** Salinity tolerance of the tree legumes: Mesquite (*Prosopis glandulosa* var. *torreyana*, *P. velutina* and *P. articulata*), algarrobo (*P. chilensis*), kiawe (*P. pallida*) and tamarugo (*P. tamarugo*) grown in sand culture on nitrogen-free media. - Plant Soil 61: 311-317, 1981.

11620 - **FELLOWS, R.J., ZEEVAART, J.A.D.:** Comparison of ethylendiaminetetraacetate--enhanced exudation from detached and translocation from attached bean leaves. - Plant Physiol. 71: 716-718, 1983.

*11621 - **FÉRARD, G., COUDRET, A.:** Relations entre l'absorption, le transport de l'ion sulfate et les contraintes hydriques chez *Plantago lanceolata* et *Plantago maritima* soumis à des variations de potentiel osmotique. - Physiol. vég. 20: 703-709, 1982.

11622 - **FETCHER, N., STRAIN, B.R., OBERBAUER, S.F.:** Effects of light regime on the growth, leaf morphology, and water relations of seedlings of two species of tropical trees. - Oecologia 58: 314-319, 1983.

*11623 - **FIELD, C., BERRY, J.A., MOONEY, H.A.:** A portable system for measuring carbon dioxide and water vapour exchange of leaves. - Plant Cell Environ. 5: 179-186, 1982.

11624 - **FIELD, C., MOONEY, H.A.:** Leaf age and seasonal effects on light, water, and nitrogen use efficiency in a California shrub. - Oecologia 56: 348-355, 1983.

11625 - **FISCUS, E.L.:** Water transport and balance within the plant: resistance to water flow in roots. - In: TAYLOR, H.M., JORDAN, W.R., SINCLAIR, T.R. (ed.): Limitations to Efficient Water Use in Crop Production. Pp. 183-194. Amer. Soc. Agron., Inc., Crop Sci. Soc. Amer., Inc., Soil Sci. Soc. Amer., Inc., Madison 1983.

11626 - **FISCUS, E.L., KLUTE, A., KAUFMANN, M.R.:** An interpretation of some whole plant water transport phenomena. - Plant Physiol. 71: 810-817, 1983.

*11627 - **FISHER, D.G., EVERT, R.F.:** Studies on the leaf of *Amaranthus retroflexus* (*Amaranthaceae*): Morphology and anatomy. - Amer. J. Bot. 69: 1133-1147, 1982.

11628 - **FITZPATRIK, G.:** Plant growth response to water rationing in a container nursery. - HortScience 18: 187-189, 1983.

11629 - **FITZSIMONS, P.J., WEYERS, J.D.B.:** Separation and purification of protoplast types from *Commelina communis* L. leaf epidermis. - J. exp. Bot. 34: 55-66, 1983.

11630 - **FLUG, M., FANGMEIER, D.D., SANDOVAL, J.E., HUSMAN, S.:** Response of *Euphorbia lathyris* to soil moisture. - Trans. ASAE 26: 789-794, 1983.

11631 - **FORBES, J.J., PRATLEY, J.E.:** The use of desiccant, defoliant and growth regulating sprays to advance the harvest of edible dry beans (*Phaseolus vulgaris*) in Tasmania. - Aust. J. exp. Agr. anim. Husb. 23: 426-428, 1983.

*11632 - **FORK, D.C., ÖQUIST, G.:** Desiccation effects on transfer of excitation energy between the two photosystems of photosynthesis at physiological temperatures in the red alga *Porphyra perforata*. - Carnegie Inst. Washington Year Book 80: 39-43, 1981.

11633 - **FORSETH, I.N., EHLERINGER, J.R.:** Ecophysiology of two solar tracking desert winter annuals. III. Gas exchage responses to light, CO_2 and VPD in relation to long-term drought. - Oecologia 57: 344-351, 1983.

11634 - **FORSETH, I.N., EHLERINGER, J.R.:** Ecophysiology of two solar tracking desert winter annuals IV. Effects of leaf orientation on calculated daily carbon gain and water use efficiency. - Oecologia 58: 10-18, 1983.

*11635 - **FRANCOIS, L.E.:** Managing alfalfa under saline conditions. - In: Twelfth California Alfalfa Symposium. Pp. 53-57. University of California, Davis 1982.

*11636 - **FRANCOIS, L.E., WEST, D.W.:** Reduction in yield and market quality of celery caused by soil salinity. - J. Amer. Soc. hort. Sci. 107: 952-954, 1982.

11637 - **FREAS, K.E., KEMP, P.R.:** Some relationships between environmental reliability and seed dormancy in desert annual plants. - J. Ecol. 71: 211-217, 1983.

*11638 - **FREDERICK, D.J., MADGWICK, H.A.I., OLIVER, G.R.:** Wood basic density and moisture content of young *Eucalyptus regnans* grown in New Zealand. - N. Zeal. J. Forest Sci. 12: 494-500, 1982.

11639 - **FREEMAN, B.M.:** Effects of irrigation and pruning of Shiraz grapevines on subsequent red wine pigments. - Amer. J. Enol. Viticult. 34: 23-26, 1983.

11640 - **FREEMAN, B.M., KLIEWER, W.M.:** Effect of irrigation, crop level and potassium fertilization on Carignane vines. II. Grape and wine quality. - Amer. J. Enol. Viticult. 34: 197-207, 1983.

*11641 - **FREEMAN, B.M., KLIEWER, W.M., STERN, P.:** Influence of windbreaks and climatic region on diurnal fluctuation of leaf water potential, stomatal conductance, and leaf temperature of grapevines. - Amer. J. Enol. Viticult. 33: 233-236, 1982.

11642 - **FREVERT, D.K., HILL, R.W., BRAATEN, B.C.:** Estimation of FAO evapotranspiration coefficients. - J. Irrig. Drain. Eng. 109: 265-270, 1983.

11643 - **FREY, W., KÜRSCHNER, H.:** Photosyntheseweg und Zonierung von Halophyten an Salzseen in der Türkei, in Jordanien und im Iran. - Flora 173: 293-310, 1983.

*11644 - **FRIEDRICH, J.W., HUFFAKER, R.C.:** The relationship between ribulose bisphosphate carboxylase concentration and photosynthesis. - In: LYONS, J.M., VALENTINE, R.C., PHILLIPS, D.A., RAINS, D.W., HUFFAKER, R.C. (ed.): Genetic Engineering of Symbiotic Nitrogen Fixation and Conservation of Fixed Nitrogen. Pp. 305-312. Plenum Press, New York - London 1981.

11645 - **FUEHRING, H.D., FINKNER, M.D.:** Effect of folicote antitranspirant application on field grain yield of moisture-stressed corn. - Agron. J. 75: 579-582, 1983.

*11646 - **FUHRER, J., KAUR-SAWHNEY, R., SHIH, L.-M., GALSTON, A.W.:** Effects of exogenous 1,3-diaminopropane and spermidine on senescence of oat leaves. II. Inhibition of ethylene biosynthesis and possible mode of action. - Plant Physiol. 70: 1597-1600, 1982.

11647 - **FUJINO, D.W., REID, M.S., KOHL, H.C.:** The water relations of maidenhair fronds treated with silver nitrate. - Scientia Hort. 19: 349-355, 1983.

*11648 - **GÁBORČÍK, N., GÁBORČÍK, S.:** Ciele a fyziologické základy šľachtenia tráv. [Aims and physiological bases of grass breeding.] - Stud. Informace ÚVTIZ (Praha), Rost. Výroba 1981(2): 1-64, 1981. [In Slov, ab: E, R.]

11649 - **GADAL, P., PERROT-RECHENMANN, C., VIDAL, J.:** Immunocytochemical visualization of phosphoenolpyruvate carboxylase in higher plants. - Physiol. vég. 21: 1055-1062, 1983.

11650 - **GAJRI, P.R., PRIHAR, S.S.:** Effect of small irrigation amounts on the yield of wheat. - Agr. Water Manage. 6: 31-41, 1983.

11651 - **GALES, K.:** Yield variation of wheat and barley in Britain in relation to crop growth and soil conditions - a review. - J. Sci. Food. Agr. 34: 1085-1104, 1983.

*11652 - **GANCHEV, S.P., ĬOTSOVA-BAURENSKA, N.M., MESHINEV, T.A., BOYADZHIĬSKI, M.N.:** Vliyanie na svetlinniya rezhim v"rkhu rastezha i razvitieto na *Potentilla fruticosa* L. v mestnostta Beglika (Zapadni Rodopi). [The effect of the light conditions on the growth and development of *Potentilla fruticosa* L. in Beglika locality (Western Rhodopes).] - Ekologiya (Sofia) 10: 3-12, 1982. [In Bulg, ab: E, R.]

11653 - **GAO, J.-Y., ANDREWS, C.J., POMEROY, M.K.:** Interactions among flooding, freezing, and ice encasement in winter wheat. - Plant Physiol. 72: 303-307, 1983.

*11654 - **GAO(KAO), Y.-Z., ZHONG, W.:** [On the relationship between photorespiration and photosynthesis II. Effects of environmental factors on photosynthesis and its relation to photorespiration.] - Acta Phytophysiol. Sin. 8: 373-384, 1982. [In Chin, ab: E.]

11655 - **GARAY, A.F., WILHELM, W.W.:** Root system characteristics of two soybean isolines undergoing water stress conditions. - Agron. J. 75: 973-977, 1983.

11656 - **GARBUTT, K., BAZZAZ, F.A.:** Leaf demography, flower production and biomass of diploid and tetraploid populations of *Phlox drummondii* Hook. on a soil moisture gradient. - New Phytol. 93: 129-141, 1983.

11657 - **GARDNER, H.R.:** Soil properties and efficient water use: evaporation of water from bare soil. - In: TAYLOR, H.M., JORDAN, W.R., SINCLAIR, T.R. (ed.): Limitations to Efficient Water Use in Crop Production. Pp. 65-71. Amer. Soc. Agron., Inc., Crop Sci. Soc. Amer., Inc., Soil Sci. Soc. Amer., Inc., Madison 1983.

11658 - **GARDNER, W.R.:** Soil properties and efficient water use: an overview. - In: TAYLOR, H.M., JORDAN, W.R., SINCLAIR, T.R. (ed.): Limitations to Efficient Water Use in Crop Production. Pp. 45-64. Amer. Soc. Agron., Inc., Crop Sci. Soc. Amer., Inc., Soil Sci. Soc. Amer., Inc., Madison 1983.

11659 - **GARDNER, W.R., GARDNER, H.R.:** Principles of water management under drought conditions. - Agr. Water Manage. 7: 143-155, 1983. Also in: STONE, J.F., WILLIS, W.O. (ed.): Plant Production and Management under Drought Conditions. Pp. 143-155. Elsevier Scientific Publishing Company, Amsterdam 1983.

11660 - **GARREC, J.-P., VAVASSEUR, A., MICHALOWICZ, G., LAFFRAY, D.:** Stomatal movements and repartition of the elements K, Cl, Na, P, Ca, Mg and S in the stomatal complexes of *Vicia faba* and *Commelina communis*. Electron probe studies. - Z. Pflanzenphysiol. 112: 35-42, 1983.

11661 - **GARRITY, D.P., SULLIVAN, C.Y., WATTS, D.G.:** Moisture deficits and grain sorghum performance: drought stress conditioning. - Agron. J. 75: 997-1004, 1983.

*11662 - **GASCHO, G.J., SHIH, S.F.:** Sugarcane. - In: TEARE, I.D., PEET, M.M. (ed.): Crop-Water Relations. Pp. 445-480. John Wiley & Sons, New York - Chichester - Brisbane - Toronto - Singapore 1982.

11663 - **GASKELL, M.L., PEARCE, R.B.:** Stomatal frequency and stomatal resistance of maize hybrids differing in photosynthetic capability. - Crop Sci. 23: 176-177, 1983.

11664 - **GATES, P., SMITH, M.L., BOULTER, D.:** Reproductive physiology of *Vicia faba* L. - In: HEBBLETHWAITE, P.D. (ed.): The Faba Bean (*Vicia faba* L.). A Basis for Improvement. Pp. 133-142. Butterworths, London - Boston - Durban - Singapore - Sydney - Toronto - Wellington 1983.

11665 - **GAY, A.P.:** Transit time diffusion porometer calibration: An analysis taking into account temperature differences and calibration non-linearity. - J. exp. Bot. 34: 461-469, 1983.

11666 - **GEBEYEHOU, G., KNOTT, D.R.:** Response of durum wheat cultivars to water stress in the field and greenhouse. - Can. J. Plant Sci. 63: 801-814, 1983.

11667 - **GEDALOVICH, E., FAHN, A.:** Ultrastructure and development of the inactive stomata of *Anabasis articulata* (Forsk.) Moq. - Amer. J. Bot. 70: 88-96, 1983.

*11668 - **GENKEL', P.A.:** Fiziologiya Zharo- i ZasukhoustoĭchivostI Rasteniĭ. [Physiology of Heat- and Drought-Resistance of Plants.] - Nauka, Moskva 1982. [In R.]

11669 - **GENSLER, W., DIAZ-MUNOZ, F.:** Stem diameter variations in cotton under field conditions. - Crop Sci. 23: 907-912, 1983.

11670 - **GENSLER, W., DIAZ-MUNOZ, F.:** Simultaneous stem diameter expansions and apoplastic electropotential variations following irrigation or rainfall in cotton. - Crop Sci. 23: 920-924, 1983.

11671 - **GEORGE, J.B., MARRIOTT, J.:** The effect of humidity in plantain ripening. - Scientia Hort. 21: 37-43, 1983.

*11672 - **GEORGE, M.F., BECWAR, M.R., BURKE, M.J.:** Freezing avoidance by deep undercooling of tissue water in winter-hardy plants. - Cryobiology 19: 628-639, 1982.

*11673 - **GEPSTEIN, S., ILAN, I.:** Does osmotic regulation of amylase activity in bean cotyledons exist? - Physiol. Plant. 53: 188-190, 1981.

11674 - **GEPSTEIN, S., JACOBS, M., TAIZ, L.:** Inhibition of stomatal opening in *Vicia faba* epidermal tissue by vanadate and abscisic acid. - Plant Sci. Lett. 28: 63-72, 1982/1983.

*11675 - **GERGELY, I., KORCAK, R.F., FAUST, M.:** Polyethylene glycol induced water stress effects on apple seedlings. I. Methodology, water consumption, and dry matter production. - J. Amer. Soc. hort. Sci. 105: 854-857, 1980.

*11676 - **GERGELY, I., KORCAK, R.F., FAUST, M.:** Polyethylene glycol induced water stress effect on apple seedlings. II. ^{45}Calcium uptake. - J. Amer. Soc. hort. Sci. 105: 858-861, 1980.

11677 - **GHUMAN, B.S., LAL, R.:** Growth and plant-water relations of sweet potato (*Ipomea batata*) as affected by soil moisture regimes. - Plant Soil 70: 95-106, 1983.

11678 - **GHUMAN, B.S., LAL, R.:** Mulch and irrigation effects on plant-water relations and performance of cassava and sweet potato. - Field Crops Res. 7: 13-29, 1983.

11679 - **GHUMAN, B.S., LAL, R.:** Effect of crop cover on temperature regime of an alfisol in the tropics. - Agron. J. 75: 931-936, 1983.

11680 - **GIAQUINTA, R.T.:** Phloem loading of sucrose. - Annu. Rev. Plant Physiol. 34: 347-387, 1983.

11681 - **GIBBS, M., BERKOWITZ, G., CHEN, C.:** Water deficit effects on non-stomatal mediated photosynthesis. - In: Sixth International Congress on Photosynthesis. Abstracts. Volume 1. Pp. 11. Brussels 1983.

11682 - **GIBERSON, W., MOTYER, M.S., GRAHAM, M.E.D., HAYDEN, D.B., THOMPSON, R.G., FENSON, D.S.:** The effect of epidermal wounds on the uptake and transport of solution fed to stems and petioles of plants. - J. exp. Bot. 34: 163-176, 1983.

*11683 - **GIFFORD, H.H., WHITEHEAD, D., THOMAS, R.S., JACKSON, D.S.:** Design of a new weighing lysimeter for measuring water use by individual trees. - N. Zeal. J. Forest. Sci. 12: 448-456, 1982.

11684 - **GILMOUR, D.J., HIPKINS, M.F., BONEY, A.D.:** The effect of osmotic stress on photosynthesis in *Dunaliella tertiolecta*. - In: Sixth International Congress on Photosynthesis. Abstracts. Volume 1. Pp. 383. Brussels 1983.

*11685 - **GINTER, G.L., COX, G.S.:** Water potential gradients in white oak, black walnut and eastern redcedar. - Trans. Missouri Acad. Sci. 14: 178, 1980.

11686 - **GINTER-WHITEHOUSE, D.L., HINCKLEY, T.M., PALLARDY, S.G.:** Spatial and temporal aspects of water relations of three tree species with different vascular anatomy. - Forest Sci. 29: 317-329, 1983.

11687 - **GIRAUD, G.:** Culture de *Fucus vesiculosus* en vue de l'expérimentation. Essais sur d'autres *Fucales*. - Physiol. vég. 21: 551-562, 1983.

11688 - **GIRI, G., SINGH, R.R., DE, R.:** Grain yield and moisture-use efficiency of dryland wheat as affected by mulch and transpiration suppressants. - Ind. J. agr. Sci. 53: 899-904, 1983.

11689 - **GISLERØD, H.R., ADAMS, P.:** Diurnal variations in the oxygen content and acid requirement of recirculating nutrient solutions and in the uptake of water and potassium by cucumber and tomato plants. - Scientia Hort. 21: 311-321, 1983.

11690 - **GLASS, A.D.M., SIDDIQI, M.Y., DEANE-DRUMMOND, C.E.:** A multichannel microcomputer-based system for continuously measuring and recording ion activities of uptake solutions during ion absorption by roots of intact plants. Plant Cell Environ. 6: 247-253, 1983.

11691 - **GLATZEL, G.:** Mineral nutrition and water relations of hemiparasitic mistletoes: a question of partitioning. Experiments with *Loranthus europaeus* on *Quercus petraea* and *Quercus robur*. - Oecologia 56: 193-201, 1983.

11692 - **GLINKA, Z., ABIR, N.:** Effect of abscisic acid on exudation from deficient and aged sunflower roots. - Physiol. Plant. 59: 208-212, 1983.

11693 - **GOLD, H.J., RAPER, C.D., Jr.:** Systems analysis and modeling in extrapolation of controlled environment studies to field conditions. - In: RAPER, C.D., Jr., KRAMER, P.J. (ed.): Crop Reactions to Water and Temperature Stresses in Humid, Temperate Climates. Pp. 315-325. Westview Press, Boulder 1983.

11694 - **GOLDOVSKIĬ, A.M., SMOLYAK, Zh.V.:** Granitsy soderzhaniya vody pri raznykh sostoyaniyakh semyan. [Limits of water content at various states of seeds.] - Fiziol. Biokhim. kul't. Rast. 15: 471-476, 1983. [In R, ab: E.]

11695 - **GOMES FILHO, E., PRISCO, J.T., CAMPOS, F.A.P., ENEAS FILHO, J.:** Effects of NaCl salinity in vivo and in vitro on ribonuclease activity of *Vigna unguiculata* cotyledons during germination. - Physiol. Plant. 59: 183-188, 1983.

11696 - **GORA, A., SCHWARZ, K.:** Ertragsbeeinflussende Wirkung verschiedener Wasserregulierungsverfahren auf Staunässeböden über altpleistozänen Deckschichten. - Arch. Acker- Pflanzenbau Bodenk. 27: 99-106, 1983.

11697 - **GOROSHKO, V.V., OBRINSKIĬ, L.E., ROZENBERG, A.M.:** Ustanovka dlya izucheniya fotosinteza, dykhaniya i transpiratsii tselykh rasteniĭ v reguliruemykh usloviyakh sredy. [An installation for studying photosynthesis, respiration and transpiration of intact plants in the controlled environment.] - Fiziol. Biokhim. kul't. Rast. 15: 392-395, 1983. [In R, ab: E.]

*11698 - **GOYAL, M.R., NELSON, G.L., DREW, L.O., CARPENTER, T.G., LOGAN, T.J.:** Moisture and soybean seedling emergence. - Trans. ASAE 24: 1432-1435, 1981.

11699 - **GRANGE, R.I., LOACH, K.:** Environmental factors affecting water loss from leafy cuttings in different propagation systems. - J. hort. Sci. 58: 1-7, 1983.

11700 - **GRANGE, R.I., LOACH, K.:** The water economy of unrooted leafy cuttings. - J. hort. Sci. 58: 9-17, 1983.

*11701 - **GREACEN, E.L., GARDNER, E.A.:** Crop behaviour on clay soils. - Trop. Agr. 59: 123-132, 1982.

11702 - **GREEN, C.F., VAIDYANATHAN, L.V., HOUGH, M.N.:** An analysis of the relationship between potential evaporation and dry-matter accumulation for winter wheat. - J. agr. Sci. 100: 351-358, 1983.

11703 - **GREEN, D.G., READ, D.W.L.:** Water use efficiency of corn, sunflower and wheat with limiting soil moisture. - Can. J. Plant Sci. 63: 747-749, 1983.

*11704 - **GREEN, T.G.A., SNELGAR, W.P.:** Carbon dioxide exchange in lichens: Relationship between the diffusive resistance of carbon dioxide and water vapour. - Lichenologist 14: 255-260, 1982.

*11705 - **GREEN, T.G.A., SNELGAR, W.P.:** A comparison of photosynthesis in two thalloid liverworts. - Oecologia 54: 275-280, 1982.

11706 - **GREENWAY, H., MUNNS, R.:** Interactions between growth, uptake of Cl^- and Na^+, and water relations of plants in saline environments. II. Highly vacuolated cells. - Plant Cell Environ. 6: 575-589, 1983.

11707 - **GREENWAY, H., MUNNS, R., WOLFE, J.:** Interactions between growth, Cl^- and Na^+ uptake, and water relations of plants in saline environments. I. Slightly vacuolated cells. - Plant Cell Environ. 6: 567-574, 1983.

*11708 - **GRIFFIN, D.M.:** Water potential as selective factor in the microbial ecology of soils. - In: Water Potential Relations of Soil Microbiology. Pp. 141-151. Soil Science Society of America, Madison 1980.

11709 - **GROUZIS, M., MÉTHY, M.:** Détermination radiométrique de la phytomasse herbacée en milieu sahélien: perspectives et limites. - Acta oecol. - Oecol. Plant. 4: 241-257, 1983.

11710 - **GUADALUPE-LUNA, R., GOYAL, M.R., CINTRÓN, M., RIVERA, L.E., De LÓPEZ, M.C.P.:** Effects of water application rates, plastic-mulch, and staking on size arrangements of mature green tomatoes under drip irrigation. - J. agr. Univ. Puerto Rico 67: 293-302, 1983.

11711 - **GUJRATHI, B.G., HEGDE, B.A., PATIL, T.M.:** Relative efficiency of photosynthetic carboxylation and enzyme activities in sorghum genotypes and peanut under water stress. - In: Sixth International Congress on Photosynthesis. Abstracts. Volume 1. Pp. 385. Brussels 1983.

11712 - **GUNSTON, H., BATCHELOR, C.H.:** A comparison of the Priestley-Taylor and Penman methods for estimating reference crop evapotranspiration in tropical countries. - Agr. Water Manage. 6: 65-77, 1983.

*11713 - **GÜNTHARDT, M.S., WANNER, H.:** Veränderungen der Spaltöffnungen und der Wachsstruktur mit zunehmendem Nadelalter bei *Pinus cembra* L. und *Picea abies* (L.) Karsten an der Waldgrenze. - Bot. Helvetica 92: 47-60, 1982.

11714 - **GUPTA, J.P., GUPTA, G.N.:** Effect of grass mulching on growth and yield of legumes. - Agr. Water Manage. 6: 375-383, 1983.

11715 - **GUPTA, J.P., SINGH, S.D.:** Hydrothermal environment of soil, and vegetable production with drip and furrow irrigations. - Ind. J. agr. Sci. 53: 138-142, 1983.

11716 - **GUSEV, N.A., PETINOV, N.S.:** Vodoobmen rasteniĭ v usloviyakh oroshaemogo zemledeliya. [Water exchange of plants under conditions of irrigation.] - Fiziol. Biokhim. kul't. Rast. 15: 15-23, 1983. [In R, ab: E.]

11717 - **HABERMANN, H.M., SHOEMAKER, E.W.:** Copper deficiency in *Helianthus annuus* L.: developmental and metabolic consequences, stomatal function and senescence. - Growth 47: 184-200, 1983.

*11718 - **HABY, V.A., BLACK, A.L., BERGMAN, J.W., LARSON, R.A.:** Nitrogen fertilizer requirements of irrigated safflower in the Northern Great Plains. - Agron. J. 74: 331-335, 1982.

*11719 - **HADLEY, N.F., SZAREK, S.R.:** Productivity of desert ecosystems. - BioScience 31: 747-753, 1981.

11720 - **HAJIBAGHERI, M.A., HALL, J.L., FLOWERS, T.J.:** The structure of the cuticle in relation to cuticular transpiration in leaves of the halophyte *Suaeda maritima* (L.) Dum. - New Phytol. 94: 125-131, 1983.

11721 - **HÁK, R., NÁTR, L.:** Energetická bilance a optické vlastnosti primárního listu jarního ječmene - vliv deficitu dusíku a fosforu. [Energy balance and optical properties of spring barley primary leaf - effect of nitrogen and phosphorus deficiency.] - In: Dny Rostlinné Fyziologie III. Pp. 191-192. ČSBS, Praha 1983. [In Czech.]

*11722 - **HALTERLEIN, A.J.:** Bean. - In: TEARE, I.D., PEET, M.M. (ed.): Crop-Water Relations. Pp. 157-186. John Wiley & Sons, New York - Chichester - Brisbane - Toronto - Singapore 1982.

11723 - **HAMER, P.J.C.:** Evaporative cooling of apple buds: the effect of timing of water application on bud development and frost resistance of the cv Cox's Orange Pippin. - J. hort. Sci. 58: 153-159, 1983.

*11724 - **HAMMOND, R.B., PEDIGO, L.P.:** Effects of artificial and insect defoliation on water loss from excised soybean leaves. - J. Kansas Entomol. Soc. 54: 331-336, 1981.

*11725 - **HAMZE, M., NIMAH, M.:** Iron content during lime-induced chlorosis with two citrus rootstocks. - J. Plant Nutr. 5: 797-804, 1982.

11726 - **HANCOCK, J.G.:** Changes in water relations and cell permeability in relation to ageing of excised squash hypocotyls: a reappraisal. - Can. J. Bot. 61: 1307-1309, 1983.

11727 - **HANDA, A.K., BRESSAN, R.A., HANDA, S., HASEGAWA, P.M.:** Clonal variation for tolerance to polyethylene glycol-induced water stress in cultured tomato cells. - Plant Physiol. 72: 645-653, 1983.

11728 - **HANDA, S., BRESSAN, R.A., HANDA, A.K., CARPITA, N.C., HASEGAWA, P.M.:** Solutes contributing to osmotic adjustment in cultured plant cells adapted to water stress. - Plant Physiol. 73: 834-843, 1983.

*11729 - **HANGARTER, R.P., GOOD, N.E.:** Energy thresholds for ATP synthesis in chloroplasts. - Biochim. biophys. Acta 681: 397-404, 1982.

11730 - **HANKS, R.J.:** Yield and water-use relationships: an overview. - In: TAYLOR, H.M., JORDAN, W.R., SINCLAIR, T.R. (ed.): Limitations to Efficient Water Use in Crop Production. Pp. 393-411. Amer. Soc. Agron., Inc., Crop Sci. Soc. Amer., Inc., Soil Sci. Soc. Amer., Inc., Madison 1983.

*11731 - **HANKS, R.J., RASMUSSEN, V.P.:** Predicting crop production as related to plant water stress. - Adv. Agron. 35: 193-215, 1982.

11732 - **HANSON, A.D., HITZ, W.D.:** Whole-plant response to water deficits: water deficits and the nitrogen economy. - In: TAYLOR, H.M., JORDAN, W.R., SINCLAIR, T.R. (ed.): Limitations to Efficient Water Use in Crop Production. Pp. 331-343. Amer. Soc. Agron., Inc., Crop Sci. Soc. Amer., Inc., Soil Sci. Soc. Amer., Inc., Madison 1983.

11733 - **HANSON, A.D., RHODES, D.:** ^{14}C tracer evidence for synthesis of choline and betaine via phosphoryl base intermediates in salinized sugarbeet leaves. - Plant Physiol. 71: 692-700, 1983.

*11734 - **HARAZONO, Y., YABUKI, K.:** [Studies on the effect of leaf-boundary layer resistance on the matter production of crop. (1) Effects of wind direction to leaf and angle of attack of air to leaf on the boundary layer resistance of sweet potato leaf.] - J. agr. Meteorol. 37: 103-110, 1981. [In Jap, ab: E.]

*11735 - **HARDJOAMIDJOJO, S., SKAGGS, R.W.:** Predicting the effects of drainage systems on corn yields. - Agr. Water Manage. 5: 127-144, 1982.

11736 - **HARIVANDI, M.A., BUTLER, J.D., SOLTANPOUR, P.N.:** Effects of soluble salts on ion accumulation in *Puccinellia* spp. - J. Plant Nutr. 6: 255-266, 1983.

*11737 - **HARMAN, W.L., UNGER, P.W., JONES, O.R.:** Sunflower yield response to furrow irrigation on fine-textured soils in the Texas High Plains. - Texas agr. Exp. Sta. MP 1521: 1-12, 1982.

11738 - **HARRIS, C.E., DHANOA, M.S.:** A non-destructive method of measuring the water content of inflorescence-bearing grass tillers. - Ann. Bot. 51: 265-268, 1983.

11739 - **HARRIS, G.C., CHEESBROUGH, J.K., WALKER, D.A.:** Measurement of CO_2 and H_2O vapor exchange in spinach leaf discs. Effects of orthophosphate. - Plant Physiol. 71: 102-107, 1983.

11740 - **HARRIS, G.C., CHEESBROUGH, J.K., WALKER, D.A.:** Effects of mannose on photosynthetic gas exchange in spinach leaf discs. - Plant Physiol. 71: 108-111, 1983.

11741 - **HARRIS, J.G., MANNERS, J.G.:** Influence of relative humidity on germination and disease development in *Erysiphe graminis*. - Trans. Brit. mycol. Soc. 81: 605-611, 1983.

11742 - **HARRISON, J.G.:** Growth of lesions caused by *Botrytis fabae* on field bean leaves in relation to foliar bacteria, non-enzymic phytotoxins, pectic enzymes and osmotica. - Ann. Bot. 52: 823-838, 1983.

11743 - **HARTUNG, W.:** The site of action of abscisic acid at the guard cell plasmalemma of *Valerianella locusta*. - Plant Cell Environ. 6: 427-428, 1983.

11744 - **HARTUNG, W., KAISER, W.M., BURSCHKA, C.:** Release of abscisic acid from leaf strips under osmotic stress. - Z. Pflanzenphysiol. 112: 131-138, 1983.

11745 - **HÄSLER, R.:** Net photosynthesis of *Pinus montana* under water stress conditions at alpine timberline. - In: Sixth International Congress on Photosynthesis. Abstracts. Volume 1. Pp. 386. Brussels 1983.

11746 - **HASSON, E., POLJAKOFF-MAYBER, A.:** Changes in osmolarity and solute content of pea plants exposed to salinity and abscisic acid. - Aust. J. Plant Physiol. 10: 573-583, 1983.

11747 - **HASSON, E., POLJAKOFF-MAYBER, A., GALE, J.:** The effect of salt species and concentration on photosynthesis and growth of pea plants (*Pisum sativum* L. cv. Alaska). - In: MARCELLE, R., CLIJSTERS, H., VAN POUCKE, M. (ed.): Effects of Stress on Photosynthesis. Pp. 305-311. Martinus Nijhoff / Dr W. Junk Publishers, The Hague - Boston - London 1983.

11748 - **HATFIELD, J.L.:** The utilization of thermal infrared radiation measurements from grain sorghum crops as a method of assessing their irrigation requirements. - Irrig. Sci. 3: 259-268, 1983.

11749 - **HATFIELD, J.L., PERRIER, A., JACKSON, R.D.:** Estimation of evapotranspiration at one time-of-day using remotely sensed surface temperatures. - Agr. Water Manage. 7: 341-350, 1983. Also in: STONE, J.F., WILLIS, W.O. (ed.): Plant Production and Management under Drought Conditions. Pp. 341-350. Elsevier Scientific Publishing Company, Amsterdam 1983.

11750 - **HAUN, J.R., COSTON, D.C.:** Relationship of daily growth and development of peach leaves and fruit to environmental factors. - J. Amer. Soc. hort. Sci. 108: 666-671, 1983.

11751 - **HAVAUX, M., CLEANIS, C., LANNOYE, R.:** Effects of low temperatures on barley chloroplast membranes. - In: MARCELLE, R., CLIJSTERS, H., VAN POUCKE, M. (ed.): Effects of Stress on Photosynthesis. Pp. 267-276. Martinus Nijhoff / Dr W. Junk Publishers, The Hague - Boston - London 1983.

11752 - **HAVAUX, M., LANNOYE, R.:** Chlorophyll fluorescence induction: A sensitive indicator of water stress in maize plants. - Irrig. Sci. 4: 147-151, 1983.

11753 - **HEATHERLY, L.G.:** Response of soybean cultivars to irrigation of a clay soil. - Agron. J. 75: 859-864, 1983.

11754 - **HEBBLETHWAITE, P.D. (ed.):** The Faba Bean (*Vicia faba* L.). A Basis for Improvement. - Butterworths, London - Boston - Durban - Singapore - Sydney - Toronto - Wellington 1983.

11755 - **HECHT-BUCHHOLZ, C.:** Light and electron microscopic investigations of the reactions of various genotypes to nutritional disorders. - In: SARIĆ, M.R., LOUGHMAN, B.C. (ed.): Genetic Aspects of Plant Nutrition. Pp. 17-31. Martinus Nijhoff Publishers, The Hague - Boston - Lancaster 1983.

*11756 - **HEICHEL, G.H.:** Alfalfa. - In: TEARE, I.D., PEET, M.M. (ed.): Crop-Water Relations. Pp. 127-155. John Wiley & Sons, New York - Chichester - Brisbane - Toronto - Singapore 1982.

11757 - **HEICHEL, G.H.:** Crop manipulation for efficient use of water: inadvertent and intended manipulations of crop water use. - In: TAYLOR, H.M., JORDAN, W.R., SINCLAIR, T.R. (ed.): Limitations to Efficient Water Use in Crop Production. Pp. 375-380. Amer. Soc. Agron., Inc., Crop Sci. Soc. Amer., Inc., Soil Sci. Soc. Amer., Inc., Madison 1983.

11758 - **HEICHEL, G.H., TURNER, N.C.:** CO_2 assimilation of primary and regrowth foliage of red maple (*Acer rubrum* L.) and read oak (*Quercus rubra* L.): response to defoliation. - Oecologia 57: 14-19, 1983.

*11759 - **HEIKAL, M.M., AHMED, A.M., SHADDAD, M.A.:** Dry matter and mineral composition of some oil producing plants as influenced by some salinization treatments. - Phyton 20: 159-173, 1980.

11760 - **HENDRIX, D.L., PIERCE, W.S.:** Osmoregulation and membrane-mediated responses to altered water potential in plant cells. - Cryobiology 20: 466-486, 1983.

11761 - **HENSON, I.E.:** Abscisic acid and water relations of rice (*Oryza sativa* L.): Effects of drought conditioning on abscisic acid accumulation in the leaf and stomatal response. - Ann. Bot. 52: 247-255, 1983.

11762 - **HENSON, I.E.:** Abscisic acid accumulation in detached leaves of rice (*Oryza sativa* L.) in response to water stress: A correlation with leaf size. - Ann. Bot. 52: 385-398, 1983.

11763 - **HENSON, I.E.:** Effects of light on water stress-induced accumulation of abscisic acid in leaves and seedling shoots of pearl millet (*Pennisetum americanum* (L.) Leeke). - Z. Pflanzenphysiol. 112: 257-268, 1983.

11764 - **HENSON, I.E., ALAGARSWAMY, G., MAHALAKSHMI, V., BIDINGER, F.R.:** Stomatal response to water stress and its relationship to bulk leaf water status and osmotic adjustment in pearl millet (*Pennisetum americanum* (L.) Leeke). - J. exp. Bot. 34: 442-450, 1983.

11765 - **HENSON, I.E., MAHALAKSHMI, V., ALAGARSWAMY, G., BIDINGER, F.R.:** An association between flowering and reduced stomatal sensitivity to water stress in pearl millet (*Pennisetum americanum* (L.) Leeke). - Ann. Bot. 52: 641-648, 1983.

11766 - **HERBERT, S.J., BAGGERMAN, F.D.:** Cowpea response to row width, density, and irrigation. - Agron. J. 75: 982-986, 1983.

11767 - **HESSE, N., LENZ, F.:** Wasserbedarf von Stangenbohnen (*Phaseolus vulgaris* L. var *vulgaris*) in Abhängigkeit von Wachstumsstadium und Klima. - Gartenbauwissenschaft 48: 74-80, 1983.

11768 - **HEUER, B., PLAUT, Z.:** Photosynthetic activity of sugarbeet cultivars varying in their salinity tolerance. - In: Sixth International Congress on Photosynthesis. Abstracts. Volume 1. Pp. 389. Brussels 1983.

11769 - **HILER, E.A., HOWELL, T.A.:** Irrigation options to avoid critical stress: an overview. - In: TAYLOR, H.M., JORDAN, W.R., SINCLAIR, T.R. (ed.): Limitations to Efficient Water Use in Crop Production. Pp. 497-517. Amer. Soc. Agron., Inc., Crop Sci. Soc. Amer., Inc., Soil Sci. Soc. Amer., Inc., Madison 1983.

11770 - **HILL, R.A., BLANKENSHIP, P.D., COLE, R.J., SANDERS, T.H.:** Effects of soil moisture and temperature on preharvest invasion of peanuts by the *Aspergillus flavus* group and subsequent aflatoxin development. - Appl. environ. Microbiol. 45: 628-633, 1983.

11771 - **HINCKLEY, T.M., DUHME, F., HINCKLEY, A.R., RICHTER, H.:** Drought relations of shrub species: assessment of the mechanisms of drought resistance. - Oecologia 59: 344-350, 1983.

11772 - **HIRAI, G.-I., TAKAHASHI, M., SHIMAMURA, N., UENO, E.:** Studies on the effects of relative humidity of the atmosphere upon the growth and physiology of rice plant. II. Relations of relative humidity of the atmosphere to the dry matter production of plant at the seedling stage. - Jap. J. Crop Sci. 52: 259-265, 1983.

11773 - **HIRASAWA, T., ARAKI, T., MATSUDA, E., ISHIHARA, K.:** [On exudation rate from the base of the leaf blade in rice plants.] - Jap. J. Crop Sci. 52: 574-581, 1983. [In Jap, ab: E.]

11774 - **HO, T.D.:** Biochemical mode of action of abscisic acid. - In: ADDICOTT, F.T. (ed.): Abscisic Acid. Pp. 147-169. Praeger Publishers, New York 1983.

11775 - **HOBBS, E.H., MUENDEL, H.-H.:** Water requirements of irrigated soybeans in southern Alberta. - Can. J. Plant Sci. 63: 855-860, 1983.

11776 - **HOFFMAN, G.J., JOBES, J.A., ALVES, W.J.:** Response of tall fescue to irrigation water salinity, leaching fraction, and irrigation frequency. - Agr. Water Manage. 7: 439-456, 1983.

11777 - **HOFFMAN, G.J., VAN GENUCHTEN, M.T.:** Soil properties and efficient water use: water management for salinity control. - In: TAYLOR, H.M., JORDAN, W.R., SINCLAIR, T.R. (ed.): Limitations to Efficient Water Use in Crop Production. Pp. 73-85. Amer. Soc. Agron., Inc., Crop Sci. Soc. Amer., Inc., Soil Sci. Soc. Amer., Inc., Madison 1983.

*11778 - **HOFFMANN, D.:** Water quantitative analysis of corn grains. Study of a new heat treatment on whole grain. - In: INGLETT, G.E. (ed.): Maize: Recent Progress in Chemistry and Technology. Pp. 45-51. Academic Press, New York 1982.

11779 - **HOLDER, G.D., GUMBS, F.A.:** Effects of irrigation on the growth and yield of banana. - Trop. Agr. 60: 25-30, 1983.

11780 - **HOLDER, G.D., GUMBS, F.A.:** Effects of waterlogging on the growth and yield of banana. - Trop. Agr. 60: 111-116, 1983.

11781 - **HOLDER, G.D., GUMBS, F.A.:** Effects of nitrogen and irrigation on the growth and yield of banana. - Trop. Agr. 60: 179-183, 1983.

*11782 - **HOLMES, R.W., MAHALL, B.E.:** Preliminary observations on the effects of flooding and desiccation upon the net photosynthetic rates of high intertidal estuarine sediments. - Limnol. Oceanogr. 27: 954-958, 1982.

11783 - **HOLTUM, J.A.M., O'LEARY, M.H., OSMOND, C.B.:** Effect of varying CO_2 partial pressure on photosynthesis and on carbon isotope composition of carbon-4 of malate from the Crassulacean acid metabolism plant *Kalanchoë daigremontiana* Hamet et Perr. - Plant Physiol. 71: 602-609, 1983.

11784 - **HOMMELS, C.H., STERK, A.A., TANCZOS, O.G.:** Genetic differentiation in *Taraxacum* and its relation to mineral nutrition. - In: SARIĆ, M.R., LOUGHMAN, B.C. (ed.): Genetic Aspects of Plant Nutrition. Pp. 229-235. Martinus Nijhoff Publishers, The Hague - Boston - Lancaster 1983.

11785 - **HOOKER, M.L., MOHIUDDIN, S.H., KANEMASU, E.T.:** The effect of irrigation timing on yield and yield components of winter wheat. - Can. J. Plant Sci. 63: 815-823, 1983.

11786 - **HOQUE, E., DATHE, W., TESCHE, M., SEMBDNER, G.:** Abscisic acid and its β-D-glucopyranosyl ester in saplings of Scots pine (*Pinus sylvestris* L.) in relation to water stress. - Biochem. Physiol. Pflanz. 178: 287-295, 1983.

11787 - **HOROWITZ, M., TAYLORSON, R.B.:** Effect of high temperatures on imbibition, germination, and thermal death of velvetleaf (*Abutilon theophrasti*) seeds. - Can. J. Bot. 61: 2269-2276, 1983.

*11788 - **HORTON, B.D., WEHUNT, E.J., EDWARDS, J.H., BRUCE, R.R., CHESNEE, J.L.:** The effects of drip irrigation and soil fumigation on 'Redglobe' peach yield and growth. - J. Amer. Soc. hort. Sci. 106: 438-443, 1981.

*11789 - **HORTON, R.F., WOODROW, L., BOESEL, I., GRODZINSKI, B.:** Light, carbon dioxide and ethylene metabolism in photosynthetic tissue. - In: JACKSON, M.B., GROUT, B., MacKENZIE, I.A. (ed.): Growth Regulators in Plant Senescence. Pp. 93-101. British Plant Growth Regulator Group, Wantage 1982.

11790 - **HSIAO, T.C., BRADFORD, K.J.:** Physiological consequences of cellular water deficits. - In: TAYLOR, H.M., JORDAN, W.R., SINCLAIR, T.R.(ed.): Limitations to Efficient Water Use in Crop Production. Pp. 227-265. Amer. Soc. Agron., Inc., Crop Sci. Soc. Amer., Inc., Soil Sci. Soc. Amer., Inc., Madison 1983.

11791 - **HSU, K.H., KIM, C.J., WILSON, L.A.:** Factors affecting water uptake of soybeans during soaking. - Cereal. Chem. 60: 208-211, 1983.

*11792 - **HUBAC, C., GUERRIER, D., VIEIRA da SILVA, J.:** Étude de la transpiration du Cotonnier (*Gossypium hirsutum* L.) en relation avec la photopériode. - Acta oecol. - Oecol. Plant. 3: 279-289, 1982.

*11793 - **HUBAC, C., VIEIRA da SILVA, J.:** Indicateurs métaboliques de contraintes mésologiques. - Physiol. vég. 18: 45-53, 1980.

11794 - **HUBÍK, E.:** Vliv meteorologických faktorů na výnos a výnosové složky jarního ječmene v intensivní řepařské oblasti. [The influence of meteorological factors on the yield and yield components in spring barley grown in an intensive beet-growing region.] - Rost. Výroba (Praha) 29: 753-760, 1983. [In Czech, ab: R, E.]

11795 - **HUCK, M.G., HILLEL, D.:** A model of root growth and water uptake accounting for photosynthesis, respiration, transpiration, and soil hydraulics. - Adv. Irrig. 2: 273-333, 1983.

11796 - **HUCK, M.G., ISHIHARA, K., PETERSON, C.M., USHIJIMA, T.:** Soybean adaptation to water stress at selected stages of growth. - Plant Physiol. 73: 422-427, 1983.

*11797 - **HUCK, M.G., TAYLOR, H.M.:** The rhizotron as a tool for root research. - Adv. Agron. 35: 1-35, 1982.

11798 - **HUDSON, S., TRAIL, J., SIMMONS, D., BALDWIN, L., MYERS, R., TOM, B., MOHR, J., TALLMAN, G.:** Preparation of enzymatically cleaned leaf epidermal strips of *Nicotiana glauca*. - Plant Sci. Lett. 32: 1-8, 1983.

*11799 - **HUFF, A., ABDEL-BAR, M.Z., RODNEY, D.R., ROTH, R.L., GARDNER, B.R.:** Enhancement of citrus regreening and peel lycopene by trickle irrigation. - HortScience 16: 301-302, 1981.

11800 - **HUGHES, G., KEATINGE, J.D.H.:** Solar radiation interception, dry matter production and yield in pigeonpea (*Cajanus cajan* L. Millspaugh). - Field Crops Res. 6: 171-178, 1983.

11801 - **HUKKERI, S.B., SHUKLA, N.P.:** Effect of soil-moisture stress at different stages of growth on the fodder yield of 'MP Chari' sorghum. - Ind. J. agr. Sci. 53: 44-48, 1983.

11802 - **HULUGALLE, N.R., WILLATT, S.T.:** The role of soil resistance in determining water uptake by plant root systems. - Aust. J. Soil Res. 21: 571-574, 1983.

11803 - **HUNT, P.G., MATHENY, T.A., REICOSKY, D.C., WOLLUM, A.G., II, SOJKA, R.E., CAMPBELL, R.B.:** Effect of irrigation and *Rhizobium japonicum* strain 110 upon yield and nitrogen accumulation and distribution of determinate soybeans. - Commun. Soil Sci. Plant Anal. 14: 223-238, 1983.

*11804 - **HUSS, J., KOCH, W.:** Frischezustand und Anwuchserfolg unterschiedlicher Kiefernpflanzen. - Forstwiss. Centralbl. 101: 13-24, 1982.

*11805 - **HUSSAIN, Z.:** Using highly saline irrigation water for a fodder barley crop. - J. agr. Sci. 96: 515-520, 1981.

*11806 - **HUSSEIN, J.**: A review of the use of resistance blocks for the measurement of soil water content in Zimbabwe. - Zimbabwe agr. J. 78: 63-67, 1981.

*11807 - **HUSSEIN, J.**: An investigation into methods of flood irrigation of orange trees. - Zimbabwe agr. J. 79: 73-78, 1982.

11808 - **HUSSEIN, J.**: A review of methods for determining available water capacities of soils and description of an improved method for estimating field capacity. - Zimbabwe J. agr. Res. 21: 73-87, 1983.

11809 - **HUTMACHER, R.B., KRIEG, D.R.**: Photosynthetic rate control in cotton. Stomatal and nonstomatal factors. - Plant Physiol. 73: 658-661, 1983.

11810 - **HUTTUNEN, S., LAINE, K.**: Effects of air-borne pollutants on the surface wax structure of *Pinus sylvestris* needles. - Ann. bot. Fenn. 20: 79-86, 1983.

11811 - **HUYGENS, H., IMPENS, I., LIPS, J.**: Net CO_2 exchange rates and diffusion resistance of leaves from chicory selections during vegetative growth. - Biol. Plant. 25: 419-424, 1983.

11812 - **HUZULÁK, J., MATEJKA, F.**: Relationship between soil moisture and leaf water potential of three forest tree species. - Biol. Plant. 25: 462-467, 1983.

11813 - **HUZULÁK, J., MATEJKA, F.**: A model for the leaf water potential daily dynamics of dominant shrub species in an oak-hornbeam forest. - Biológia (Bratislava) 38: 411-418, 1983.

11814 - **HUZULÁK, J., MATEJKA, F.**: Vzťah medzi evapotranspiráciou a teplotou porastu ozimnej pšenice. [Relationship between evapotranspiration and temperature of winter wheat canopy.] - In: Dny Rostlinné Fyziologie III. Pp. 203-204. ČSBS, Praha 1983. [In Slov.]

*11815 - **IBRAHIM, M., RAPP, M., LOSSAINT, P.**: Economie de l'eau d'un écosystème à *Pinus pinea* L. du littoral méditerranéen. - Ann. Sci. forest. 39: 289-306, 1982.

11816 - **IDLE, D.B., PROCTOR, C.W.**: An integrating sphere leaf chamber. - Plant Cell Environ. 6: 437-439, 1983.

*11817 - **IDSO, S.B.**: Humidity measurement by infrared thermometry. - Remote Sensing Environ. 12: 87-91, 1982.

11818 - **IDSO, S.B.**: Stomatal regulation of evaporation from well-watered plant canopies: a new synthesis. - Agr. Meteorol. 29: 213-217, 1983.

11819 - **IDSO, S.B., QUINN, J.A.**: Vegetational redistribution in Arizona and New Mexico in response to a doubling of the atmospheric CO_2 concentration. - Climatol. Publ. 17: 1-52, 1983.

11820 - **IL'KUN, G.M., YUDIN, Yu.N., KUSTOVSKIĬ, S.E.**: Sorbtsiya polyarnykh gazov list'yami drevesnykh rasteniĭ. [Sorption of polar gases by woody plant leaves.] - Fiziol. Biokhim. kul't. Rast. 15: 482-487, 1983. [In R, ab: E.]

11821 - **IMAMUL HUQ, S.M., LARHER, F.**: Osmoregulation in higher plants: effects of NaCl salinity on non-nodulated *Phaseolus aureus* L. I. Growth and mineral content. - New Phytol. 93: 203-208, 1983.

11822 - **IMAMUL HUQ, S.M., LARHER, F.**: Osmoregulation in higher plants: effects of NaCl salinity on non-nodulated *Phaseolus aureus* L. II. Changes in organic solutes. - New Phytol. 93: 209-216, 1983.

11823 - **IMAMUL HUQ, S.M., LARHER, F.:** Effect of NaCl salinity on the growth and the nitrogen status of nodulated cowpea (*Vigna sinensis* L.) and mung bean (*Phaseolus aureus* L.). - Z. Pflanzenphysiol. 112: 79-87, 1983.

11824 - **INNES, P., BLACKWELL, R.D.:** Some effects of leaf posture on the yield and water economy of winter wheat. - J. agr. Sci. 101: 367-376, 1983.

*11825 - **IRELAND, C.R., SCHWABE, W.W.:** Studies on the role of photosynthesis in the photoperiodic induction of flowering in the short-day plants *Kalanchoe blossfeldiana* Poellniz and *Xanthium pensylvanicum* Wallr. I. The requirement for CO_2 during photoperiodic induction. - J. exp. Bot. 33: 738-747, 1982.

11826 - **IRWIN, C.C., PRICE, H.C.:** The relationship of radicle length to chilling sensitivity of pregerminated pepper seed. - J. Amer. Soc. hort. Sci. 108: 484-486, 1983.

11827 - **ISHIHARA, K., SAITO, H.:** Relationship between leaf water potential and photosynthesis in rice plants. - JARQ 17: 81-86, 1983.

11828 - **ISHIKAWA, H., ABOU, S., OHTA, E., SAKATA, M.:** Effects of periodic- and transient-osmotic stress on electric potentials in bean roots. - Plant Cell Physiol. 24: 1129-1135, 1983.

11829 - **ISHIKAWA, H., AIZAWA, H., KISHIRA, H., OGAWA, T., SAKATA, M.:** Light-induced changes of membrane potential in guard cells of *Vicia faba*. - Plant Cell Physiol. 24: 769-772, 1983.

11830 - **ITIER, B., KATERJI, N.:** Une expression simplifiée de la résistance aérodynamique d'une couvert végétal. (Application à la mesure des flux de masse et d'énergie). - Acta oecol. - Oecol. gen. 4: 215-227, 1983.

11831 - **IU, K.L., PULFORD, I.D., DUNCAN, H.J.:** Influence of soil waterlogging on subsequent plant growth and trace metal content. - Plant Soil 66: 423-427, 1982.

*11832 - **IZMAĬLOVA, N.N.:** Vodnyĭ rezhim gornykh rasteniĭ aridnykh oblasteĭ. [Water regime of mountain plants in dry regions.] - In: Ėkologiya i Biologiya Vysokogornykh Rasteniĭ. (Problemy Bot. 14(2). Pp. 23-27. Nauka, Novosibirsk 1979. [In R.]

*11833 - **IZZELDIN, H., LIPPERT, L.F., TAKATORI, F.H.:** An influence of water stress at different growth stages on yield and quality of lettuce seed. - J. Amer. Soc. hort. Sci. 105: 68-71, 1980.

11834 - **JACKSON, M.B., KOWALEWSKA, A.K.B.:** Positive and negative messages from roots induce foliar desiccation and stomatal closure in flooded pea plants. - J. exp. Bot. 34: 493-506, 1983.

11835 - **JACKSON, R.D., HATFIELD, J.L., REGINATO, R.J., IDSO, S.B., PINTER, P.J., Jr.:** Estimation of daily evapotranspiration from one time-of-day measurements. - Agr. Water Manage. 7: 351-362, 1983. Also in: STONE, J.F., WILLIS, W.O. (ed.): Plant Production and Management under Drought Conditions. Pp. 351-362. Elsevier Scientific Publishing Company, Amsterdam 1983.

11836 - **JACKSON, R.D., SLATER, P.N., PINTER, P.J., Jr.:** Discrimination of growth and water stress in wheat by various vegetation indices through clear and turbid atmospheres. - Remote Sensing Environ. 13: 187-208, 1983.

11837 - **JACOBI, W.R., MAIN, C.E., POWELL, N.T.:** Influence of temperature and rainfall on the development of tobacco black shank. - Phytopathology 73: 139-143, 1983.

11838 - **JAGELS, R.:** Further evidence for osmoregulation in epidermal leaf cells of seagrasses. - Amer. J. Bot. 70: 327-333, 1983.

11839 - **JAIN, J.C., QUICK, W.A., HSIAO, A.I.:** ATP synthesis during water imbibition in caryopses of genetically dormant and non-dormant lines of wild oat (*Avena fatua* L.). - J. exp. Bot. 34: 381-387, 1983.

*11840 - **JANJIĆ, V.:** Fiziološke osnove otpornosti *Setaria glauca* na djelovanje herbicida glifosata i cijanazina. [Physiological bases of the resistance of *Setaria glauca* to the effects of herbicides glyphosate and cyanazine.] - Fragm. herbol. Jugosl. 10(2): 45-50, 1981. [In Croat, ab: E.]

11841 - **JARVIS, P.G., LEVERENZ, J.W.:** Productivity of temperate, deciduous and evergreen forests. - In: LANGE, O.L., NOBEL, P.S., OSMOND, C.B., ZIEGLER, H. (ed.): Physiological Plant Ecology IV. Ecosystem Processes: Mineral Cycling, Productivity and Man's Influence. Pp. 233-280. Springer-Verlag, Berlin - Heidelberg - New York 1983.

11842 - **JEJE, A., ZIMMERMANN, M.:** The anisotropy of the mesophyll and CO_2 capture sites in *Vicia faba* L. leaves at low light intensities. - J. exp. Bot. 34: 1676-1694, 1983.

11843 - **JENSEN, S.D., CAVALIERI, A.J.:** Drought tolerance in US maize. - Agr. Water Manage. 7: 223-236, 1983. Also in: STONE, J.F., WILLIS, W.O. (ed.): Plant Production and Management under Drought Conditions. Pp. 223-236. Elsevier Scientific Publishing Company, Amsterdam 1983.

11844 - **JOARDER, O.I.:** Oil content, yield and morphological response of rape (*Brassica campestris* L.) to irrigation and fertilization treatment. - J. agr. Sci. 100: 253-255, 1983.

11845 - **JODARI-KARIMI, F., WATSON, V., HODGES, H., WHISLER, F.:** Root distribution and water use efficiency of alfalfa as influenced by depth of irrigation. - Agron. J. 75: 207-211, 1983.

11846 - **JOHNS, D., BEARD, J.B., Van BAVEL, C.H.M.:** Resistances to evapotranspiration from a St. Augustinegrass turf canopy. - Agron. J. 75: 419-422, 1983.

11847 - **JOHNSON, D.A., RICHARDS, R.A., TURNER, N.C.:** Yield, water relations, gas exchange, and surface reflectances of near-isogenic wheat lines differing in glaucousness. - Crop Sci. 23: 318-325, 1983.

11848 - **JOHNSON, J.D., FERRELL, W.K.:** Stomatal response to vapour pressure deficit and the effect of plant water stress. - Plant Cell Environ. 6: 451-456, 1983.

11849 - **JOHNSON, J.D., PFISTER, V.R., HOMANN, P.H.:** Metastable proton pools in thylakoids and their importance for the stability of photosystem II. - Biochim. biophys. Acta 723: 256-265, 1983.

11850 - **JOHNSON, J.D., RICHTER, M.L., HOMANN, P.H.:** Protons, salts and osmolarity as probes of the organization of the two photosystems. - In: Sixth International Congress on Photosynthesis. Abstracts. Volume 1. Pp. 293. Brussels 1983.

11851 - **JOINER, J.N., CONOVER, C.A., POOLE, R.T.:** Influence of leaf shine compounds on light compensation point of *Ficus benjamina*. - HortScience 18: 373-374, 1983.

11852 - **JOLIVET, Y., HAMELIN, J., LARHER, F.:** Osmoregulation in halophytic higher plants: The protective effects of glycine betaine and other related solutes against the oxalate destabilization of membranes in beet root cells. - Z. Pflanzenphysiol. 109: 171-180, 1983.

*11853 - **JOLY, C.A., CRAWFORD, R.M.M.:** Variation in tolerance and metabolic responses to flooding in some tropical trees. - J. exp. Bot. 33: 799-809, 1982.

11854 - **JONES, H., TOMOS, A.D., LEIGH, R.A., WYN JONES, R.G.:** Water-relation parameters of epidermal and cortical cells in the primary root of *Triticum aestivum* L. - Planta 158: 230-236, 1983.

11855 - **JONES, H.G.:** Plants and Microclimate. A Quantitative Approach to Environmental Plant Physiology. - Cambridge University Press, Cambridge - London - New York - New Rochelle - Melbourne - Sydney 1983.

11856 - **JONES, H.G., FANJUL, L.:** Effects of water stress on CO_2 exchange in apple. - In: MARCELLE, R., CLIJSTERS, H., VAN POUCKE, M. (ed.): Effects of Stress on Photosynthesis. Pp. 75-84. Martinus Nijhoff / Dr W. Junk publishers, The Hague - Boston - London 1983.

11857 - **JONES, H.G., HIGGS, K.H.:** Cell sap osmotic potentials and frost tolerance in black currants (*Ribes nigrum* L.). - J. hort. Sci. 58: 261-266, 1983.

11858 - **JONES, H.G., LUTON, M.T., HIGGS, K.H., HAMER, P.J.C.:** Experimental control of water status in an apple orchard. - J. hort. Sci. 58: 301-316, 1983.

*11859 - **JONES, J.B., Jr., PALLAS, J.E., Jr., STANSELL, J.R.:** Tracking the elemental content of leaves and other plant parts of the peanut under irrigated culture in the sandy soils of South Georgia. - Commun. Soil Sci. Plant Anal. 11: 81-92, 1980.

11860 - **JONES, J.W.:** Irrigation options to avoid critical stress: optimization of on-farm water allocation to crops. - In: TAYLOR, H.M., JORDAN, W.R., SINCLAIR, T.R. (ed.): Limitations to Efficient Water Use in Crop Production. Pp. 507-516. Amer. Soc. Agron., Inc., Crop Sci. Soc. Amer., Inc., Soil Sci. Soc. Amer., Inc., Madison 1983.

11861 - **JONES, J.W., ZUR, B., BOOTE, K.J.:** Field evaluation of a water relations model for soybean. II. Diurnal fluctuations. - Agron. J. 75: 281-286, 1983.

11862 - **JONES, V.P., YOUNGMAN, R.R., PARRELLA, M.P.:** Effect of selected acaricides on photosynthetic rates of lemon and orange leaves in California. - J. econ. Entomol. 76: 1178-1180, 1983.

*11863 - **JORDAN, P.W., NOBEL, P.S.:** Height distributions of two species of cacti in relation to rainfall, seedlings establishment, and growth. - Bot. Gaz. 143: 511-517, 1982.

*11864 - **JORDAN, W.R.:** Cotton. - In: TEARE, I.D., PEET, M.M. (ed.): Crop-Water Relations. Pp. 213-254. John Wiley & Sons, New York - Chichester - Brisbane - Toronto - Singapore 1982.

11865 - **JORDAN, W.R.:** Whole plant response to water deficits: an overview. - In: TAYLOR, H.M., JORDAN, W.R., SINCLAIR, T.R. (ed.): Limitaions to Efficient Water Use in Crop Production. Pp. 289-317. Amer. Soc. Agron., Inc., Crop Sci. Soc. Amer., Inc., Soil Sci. Soc. Amer., Inc., Madison 1983.

11866 - **JOSIS, P., NDAYISHIMIYE, V., RENARD, C.:** Étude des relations hydriques chez *Coffea arabica* L. II. Évaluation de la résistance à la sécheresse de divers cultivars à Gisha (Burundi). - Café Cacao Thé 27: 275-282, 1983.

*11867 - **JÖSSANG, P., CARBONNIER, J., MONFILS, M., MOLHO, D.:** Highly active ionophoric transpiration inhibitors in cut barley leaves. - Naturwissenschaften 67: 199, 1980.

11868 - **JOYCE, D.C., ASPINALL, D., EDWARDS, G.R.:** Water deficit and the growth and anatomy of the radish fleshy axis. - New Phytol. 93: 439-446, 1983.

11869 - **JUNIPER, B.E., JEFFREE, C.E.:** Plant Surfaces. - Edward Arnold, London 1983.

11870 - **JURKO, A.:** Ecomorphological survey of the leaf characteristics in selected Central European plant communities. - Folia Geobot. Phytotaxon. 18: 287-300, 1983.

*11871 - **KACHEL, K., ROTH, D.:** Einfluss unterschiedlicher Beregnungssteuerungsmethoden auf Zusatzwasseraufwand, Bodenfeuchteverlauf, Ertrag und Zustazwasserausnutzung bei Zuckerrüben, Welschem Weidelgras und Sommergerste. - Arch. Acker- Pflanzenbau Bodenk. 26: 541-552, 1982.

11872 - **KAISER, W.M.:** Sites and mechanisms for the inhibition of photosynthesis under dehydration. - In: Sixth International Congress on Photosynthesis. Abstracts. Volume 1. Pp. 15. Brussels 1983.

11873 - **KAISER, W.M.:** Photosynthetic activity and osmotic volumes of isolated intact chloroplasts and of cells in leaf tissue from various plants under osmotic stress. - In: MARCELLE, R., CLIJSTERS, H., VAN POUCKE, M. (ed.): Effects of Stress on Photosynthesis. Pp. 55-63. Martinus Nijhoff / Dr W. Junk Publishers, The Hague - Boston - London 1983.

11874 - **KAISER, W.M., SCHWITULLA, M., WIRTH, E.:** Reactions in chloroplasts, cytoplasm and mitochondria of leaf slices under osmotic stress. - Planta 158: 302-308, 1983.

11875 - **KAISER, W.M., WEBER, H., SAUER, M.:** Photosynthetic capacity, osmotic response and solute content of leaves and chloroplasts from *Spinacia oleracea* under salt stress. - Z. Pflanzenphysiol. 113: 15-27, 1983.

*11876 - **KALIR, A., FLOWERS, T.J.:** The effect of salts on malate dehydrogenase from leaves of *Zea mays*. -Phytochemistry 21: 2189-2193, 1982.

11877 - **KANEMASU, E.T.:** Yield and water-use relationships: some problems of relating grain yield to transpiration. - In: TAYLOR, H.M., JORDAN, W.R., SINCLAIR, T.R. (ed.): Limitations to Efficient Water Use in Crop Production. Pp. 413-417. Amer. Soc. Agron., Inc., Crop Sci. Soc. Amer., Inc., Soil Sci. Soc. Amer., Inc., Madison 1983.

11878 - **KANEMASU, E.T., STEINER, J.L., BIERE, A.W., WORMAN, F.D., STONE, J.F.:** Irrigation in the Great Plains. - Agr. Water Manage. 7: 157-178, 1983. Also in: STONE, J.F., WILLIS, W.O. (ed.): Plant Production and Management under Drought Conditions. Pp. 157-178. Elsevier Scientific Publishing Company, Amsterdam 1983.

11879 - **KANG, M.S., ZUBER, M.S., KRAUSE, G.F.:** Path coefficient analyses of grain yield and harvest grain moisture in maize. - Trop. Agr. 60: 253-256, 1983.

11880 - **KANNANGARA, T., SEETHARAMA, N., DURLEY, R.C., SIMPSON, G.M.:** Drought resistance of *Sorghum bicolor*. 6. Changes in endogenous growth regulators of plants grown across an irrigation gradient. - Can. J. Plant Sci. 63: 147-155, 1983.

11881 - **KAPPEL, F., FLORE, J.A.:** Effect of shade on photosynthesis, specific leaf weight, leaf chlorophyll content, and morphology of young peach trees. - J. Amer. Soc. hort. Sci. 108: 541-544, 1983.

11882 - **KAPUYA, J.A., HOZA, J.T.:** Fine structure of the abscission layer in *Phaseolus* flowers under water stress conditions. - Acta bot. Neerl. 32: 307-311, 1983.

*11883 - **KARADGE, B.A., CHAVAN, P.D.:** Salt tolerance studies in groundnut (*Arachis hypogaea* L.) variety TMV-10. - Biovigyanam 7: 137-144, 1981.

*11884 - **KARADGE, B.A., CHAVAN, P.D.:** Salinity and initial products of photosynthesis in peanut (*Arachis hypogaea* L.). - Biovigyanam 8: 95-96, 1982.

11885 - **KARADGE, B.A., CHAVAN, P.D.:** Physiological studies in salinity tolerance of *Sesbania aculeata* Poir. - Biol. Plant. 25: 412-418, 1983.

11886 - **KARAMANOS, A.J., DROSSOPOULOS, J.B., NIAVIS, C.A.:** Free proline accumulation during development of two wheat cultivars with water stress. - J. agr. Sci. 100: 429-439, 1983.

11887 - **KARLIC, H., RICHTER, H.:** Developmental effects on leaf water relations of two evergreen shrubs (*Prunus laurocerasus* L. and *Ilex aquifolium* L.). - Flora 173: 143-150, 1983.

11888 - **KARLSSON, P.E., HÖGLUND, H.-O., KLOCKARE, R.:** Blue light induces stomatal transpiration in wheat seedlings with chlorophyll deficiency caused by SAN-9789. - Physiol. Plant. 57: 417-421, 1983.

*11889 - **KARMARKAR, S.M.:** Senescence in mangroves. - In: SEN, G.V., RAJPUROHIT, K.S. (ed.): Contributions to the Ecology of Halophytes. Pp. 173-187. Dr W. Junk Publishers, The Hague - Boston - London 1982.

*11890 - **KARMOKER, J.L.:** The effect of kinetin on volume flow and ion flux in excised root systems of *Phaseolus vulgaris* L. cv. Redland Pioneer. - Bangladesh J. Bot. 11: 93-100, 1982.

11891 - **KATASE, M., USHIJIMA, T., TAZAKI, T.:** The relationship between absorption of sulfur dioxide (SO_2) and inhibition of photosynthesis in several plants. - Bot. Mag. (Tokyo) 96: 1-13, 1983.

11892 - **KATERJI, N., HALLAIRE, M., MENOUX-BOYER, Y., PERRIER, A.:** Transfert hydrique dans le végétal. II. - Détermination au champ des variations de contenu en eau du couvert végétal. - Acta oecol. - Oecol. Plant. 4: 171-181, 1983.

11893 - **KATERJI, N., HALLAIRE, M., PERRIER, A., DURAND, R.:** Transfert hydrique dans le végétal. I. - Modélisation à l'échelle du couvert végétal en conditions naturelles. - Acta oecol. - Oecol. Plant. 4: 11-26, 1983.

11894 - **KATERJI, N., OULID-AÏSSA, A.K.:** Étude et amélioration de la fiabilité d'un poromètre à diffusion utilisé en milieu naturel. - Agronomie 3: 931-936, 1983.

11895 - **KATERJI, N., PERRIER, A.:** Modélisation de l'évapotranspiration réelle ETR d'une parcelle de luzerne: rôle d'un coefficient cultural. - Agronomie 3: 513-521, 1983.

11896 - **KATERJI, N., PERRIER, A., OULID-AÏSSA, A.K.:** Exploration au champ et interprétation de la variation horizontale et verticale de la résistance stomatique: cas d'une culture de luzerne (*Medicago sativa* L.). - Agronomie 3: 847-856, 1983.

*11897 - **KAUFMANN, M.R., THOR, G.L.:** Measurement of water stress in subalpine trees: effects of temporary tissue storage methods and needle age. - Can. J. Forest Res. 12: 969-972, 1982.

11898 - **KAWASAKI, T., AKIBA, T., MORITSUGU, M.:** Effects of high concentrations of sodium chloride and polyethylene glycol on the growth and ion absorption in plants. I. Water culture experiments in a greenhouse. - Plant Soil 75: 75-85, 1983.

11899 - **KAWASAKI, T., SHIMIZU, G., MORITSUGU, M.:** Effects of high concentrations of sodium chloride and polyethylene glycol on the growth and ion absorption in plants. II. Multi-compartment transport box experiment with excised roots of barley. - Plant Soil 75: 87-93, 1983.

11900 - **KAZAMA, H., KATSUMI, M.:** Gibberellin-induced changes in the water absorption, osmotic potential and starch content of cucumber hypocotyls. - Plant Cell Physiol. 24: 1209-1216, 1983.

11901 - **KEATINGE, J.D.H., COOPER, P.J.M.:** Kabuli chickpea as a winter-sown crop in northern Syria: moisture relations and crop productivity. - J. agr. Sci. 100: 667-680, 1983.

*11902 - **KEATINGE, J.D.H., STEWART, R.H., GARRETT, M.K.:** The influence of temperature and soil water potential on the leaf extension rate of perennial ryegrass in Northern Ireland. - J. agr. Sci. 92: 175-183, 1979.

11903 - **KEELEY, J.E.:** Lack of diurnal acid metabolism in two terrestrial *Isoetes* species. - Photosynthetica 17: 93-94, 1983.

11904 - **KEENER, M.E., KIRCHER, P.L.:** The use of canopy temperature as an indicator of drought stress in humid regions. - Agr. Meteorol. 28: 339-349, 1983.

*11905 - **KELLEY, D.B., GOODIN, J.R., MILLER, D.R.:** Biology of *Atriplex*. - In: SEN, G.V., RAJPUROHIT, K.S. (ed.): Contributions to the Ecology of Halophytes. Pp. 79-107. Dr W. Junk Publishers, The Hague - Boston - London 1982.

*11906 - **KELLY, G.J., LATZKO, E.:** Carbon metabolism: the profound effects of illumination on the metabolism of photosynthetic cells. - Progress Bot. 44: 103-131, 1982.

11907 - **KENNEDY, C.W., BUSH, L.P.:** Effect of environmental and management factors on the accumulation of N-acetyl and N-formyl loline alcaloids in tall fescue. - Crop Sci. 23: 547-552, 1983.

11908 - **KENNEDY, R.A., VANDER ZEE, D., BOZARTH, C.S.:** Chloroplast development and photosynthesis in *Echinochloa* (barnyard grass) and rice: O_2 and flooding stress. - In: MARCELLE, R., CLIJSTERS, H., VAN POUCKE, M. (ed.): Effects of Stress on Photosynthesis. Pp. 341-351. Martinus Nijhoff / Dr W. Junk Publishers, The Hague - Boston - London 1983.

11909 - **KERSHAW, K.A.:** The thermal operating-environment of a lichen. - Lichenologist 15: 191-207, 1983.

11910 - **KERSHAW, K.A., MACFARLANE, J.D., WEBBER, M.R., FOVARGUE, A.:** Phenotypic differences in the seasonal pattern of net photosynthesis in *Cladonia stellaris*. - Can. J. Bot. 61: 2169-2180, 1983.

11911 - **KERSHAW, K.A., WATSON, S.:** The control of seasonal rates of net photosynthesis by moisture, light and temperature in *Parmelia disjuncta* Erichs. - Bryologist 86: 31-43, 1983.

*11912 - **KHADZHIKURBANOVA, G.:** Sezonnye izmeneniya osmoticheskikh pokazateleĭ polukustarnikov yugo-zapadnogo Kyzylkuma. [Seasonal changes in the osmotic indices of semi-shrubs in the southwestern Kyzyl Kum desert.] - Uzb. biol. Zh. 1980(1): 26-29, 1980. [In R.]

11913 - **KHAN, A.A., PECK, N.H., TAYLOR, A.G., SAMIMY, C.:** Osmoconditioning of beet seeds to improve emergence and yield in cold soil. - Agron. J. 75: 788-794, 1983.

*11914 - **KHAN, A.R., DATTA, B.:** Scheduling of irrigation for summer peanuts. - Peanut Sci. 9: 10-13, 1982.

11915 - **KHATRI, T.J., PATEL, R.M.:** Use of yield-prediction model for scheduling irrigation to groundnut. - Ind. J. agr. Sci. 53: 831-832, 1983.

*11916 - **KHIDSE, S.R., BHALE, N.L., BORIKAR, S.T.:** Combining ability for leaf water deficit and grain yield in rabi sorghum. - Ind. J. Genet. Plant Breed. 42: 82-86, 1982.

11917 - **KHIDSE, S.R., BHALE, N.L., BORIKAR, S.T.:** Leaf-water deficit in sorghum hybrids and their parents. - Ind. J. agr. Sci. 53: 289-291, 1983.

*11918 - **KILLILEA, D.M., McNEILL, L.:** Variation of xylem-water potential of twigs of coppicing shoots of *Brachystegia spiciformis* Benth. during the rainy season. - Zimbabwe J. agr. Res. 20: 169-178, 1982.

11919 - **KIMBALL, B.A.:** Canopy gas exchange: gas exchange with soil. - In: TAYLOR, H.M., JORDAN, W.R., SINCLAIR, T.R. (ed.): Limitations to Efficient Water Use in Crop Production. Pp. 215-226. Amer. Soc. Agron., Inc., Crop Sci. Soc. Amer., Inc., Soil Sci. Soc. Amer., Inc., Madison 1983.

11920 - **KIMBALL, B.A., IDSO, S.B.:** Increasing atmospheric CO_2: effect on crop yield, water use and climate. - Agr. Water Manage. 7: 55-72, 1983. Also in: STONE, J.F., WILLIS, W.O. (ed.): Plant Production and Management under Drought Conditions. Pp. 55-72. Elsevier Scientific Publishing Company, Amsterdam 1983.

*11921 - **KINRA, K.L., KUSHWAHA, B.L.:** Note on irrigation requirement of linseed for seed yield. - Ind. J. agr. Sci. 52: 478-480, 1982.

*11922 - **KIRKHAM, M.B.:** Temperature coefficients as a screening test for drought resistance in wheat. - Ind. J. Genet. Plant Breed. 41: 110-113, 1981.

11923 - **KIRKHAM, M.B.:** Effect of ethephon on the water status of a drought-resistant and drought-sensitive cultivar of winter wheat. - Z. Pflanzenphysiol. 112: 103-112, 1983.

11924 - **KIRKHAM, M.B.:** Physical model of water in a split-root system. - Plant Soil 75: 153-168, 1983.

11925 - **KIRKHAM, M.B.:** Effect of ABA on the water relations of winter-wheat cultivars varying in drought resistance. - Physiol. Plant. 59: 153-157, 1983.

11926 - **KIRKHAM, M.B., JOHNSON, D.E., Jr., KANEMASU, E.T., STONE, L.R.:** Canopy temperature and growth of differentially irrigated alfalfa. - Agr. Meteorol. 29: 235-246, 1983.

*11927 - **KIRKHAM, M.B., KANEMASU, E.T.:** Wheat. - In: TEARE, I.D., PEET, M.M. (ed.): Crop-Water Relations. Pp. 481-520. John Wiley & Sons, New York - Chichester - Brisbane - Toronto - Singapore 1982.

*11928 - **KIRST, G.O., KRAMER, D.:** Cytological evidence for cytoplasmic volume control in *Platymonas subcordiformis* after osmotic stress. - Plant Cell Environ. 4: 455-462, 1981.

*11929 - **KISHITANI, S., TSUNODA, S.:** Leaf thickness and response of leaf photosynthesis to water stress in soybean varieties. - Euphytica 31: 657-664, 1982.

11930 - **KISLYUK, I.M., VAS'KOVSKIĬ, M.D., BUBOLO, L.S., PALEEVA, T.V.:** Vliyanie temperatury na stroenie list'ev i fotosintez *Carex lugens* (*Cyperaceae*) i *Arctagrostis arundinacea* (*Poaceae*). [The effect of temperature on the leaf structure and photosynthesis in *Carex lugens* (*Cyperaceae*) and *Arctagrostis arundinacea* (*Poaceae*).] - Bot. Zh. 68: 1325-1332, 1983. [In R, ab: E.]

11931 - **KITTOCK, D.L., HENNEBERRY, T.J., BARIOLA, L.A., TAYLOR, B.B., HOFMANN, W.C.:** Cotton boll period response to water stress and pink bollworm. - Agron. J. 75: 17-20, 1983.

*11932 - **KLEINKOPF, G.E.:** Potato. - In: TEARE, I.D., PEET, M.M. (ed.): Crop-Water Relations. Pp. 287-305. John Wiley & Sons, New York - Chichester - Brisbane - Toronto - Singapore 1982.

11933 - **KLEPPER, B.:** Managing root systems for efficient water use: axial resistances to flow in root systems - anatomical considerations. - In: TAYLOR, H.M., JORDAN, W.R., SINCLAIR, T.R. (ed.): Limitations to Efficient Water Use in Crop Production. Pp. 115-125. Amer. Soc. Agron., Inc., Crop Sci. Soc. Amer., Inc., Soil Sci. Soc. Amer., Inc., Madison 1983.

11934 - **KLIEWER, W.M., FREEMAN, B.M., HOSSSOM, C.:** Effect of irrigation, crop level and potassium fertilization on Carignane vines. I. Degree of water stress and effect on growth and yield. - Amer. J. Enol. Viticult. 34: 186-196, 1983.

*11935 - **KNYPL, J.S.:** Osmoconditioning of soybean (*Glycine max*) seeds to improve the field peformance at early planting. - Acta Physiol. Plant. 3: 117-124, 1981.

11936 - **KOBATA, T., TAKAMI, S-I.:** [Grain production and dry matter partition in rice (*Oryza sativa* L.) in response to water deficits during the whole grain--filling period.]- Jap. J. Crop Sci. 52: 283-290, 1983.[In Jap, ab: E.]

11937 - **KOKUBUN, M., WATANABE, K.:** Analysis of the yield-determining process of field-grown soybeans in relation to canopy structure. VII. Effects of source and sink manipulations during reproductive growth on yield and yield components. - Jap. J. Crop Sci. 52: 215-219, 1983.

11938 - **KOLLÖFFEL, C., MATTHEWS, S.:** Respiratory activity in pea cotyledons during seed development. - J. exp. Bot. 34: 1026-1036, 1983.

11939 - **KONDO, K., YAGUCHI, Y.:** Stomatal responses to prey capture and trap narrowing in Venus's flytrap (*Dionaea muscipula* Ellis). II. Effects of various chemical substances on stomatal responses and trap closure. - Fyton 43: 1-8, 1983.

11940 - **KÖNIGSHOFER, H.:** Changes in ion composition and hexitol content of different *Plantago* species under the influence of salt stress. - Plant Soil 72: 289-296, 1983. Also in: SARIĆ, M.R., LOUGHMAN, B.C. (ed.): Genetic Aspects of Plant Nutrition. Pp. 243-250. Martinus Nijhoff Publishers, The Hague - Boston - Lancaster 1983.

11941 - **KONSTANTINOV, K., LAZIĆ, V., DENIĆ, M., RADENOVIĆ, Č., FURTULA, V.:** Influence of chlorate ions on some characteristics of maize seedlings with different protein content. - In: SARIĆ, M.R., LOUGHMAN, B.C. (ed.): Genetic Aspects of Plant Nutrition. Pp. 251-254. Martinus Nijhoff Publishers, The Hague - Boston - Lancaster 1983.

11942 - **KOPP, M., GEOFFROY, P., FRITIC, B.:** Plant disease and the regulation of enzymes involved in lignification. Effects of osmotic pressure on phenylpropanoid enzymes and on the hypersensitive response of tobacco mosaic virus. - Planta 157: 180-189, 1983.

11943 - **KÖRNER, C., ALLISON, A., HILSCHER, H.:** Altitudinal variation of leaf diffusive conductance and leaf anatomy in heliophytes of montane New Guinea and their interrelation with microclimate. - Flora 174: 91-135, 1983.

11944 - **KORNHER, A., TORSSELL, B.W.R.:** Simulation of weather x management interactions in temporary grasslands in Sweden. - Swedish J. agr. Res. 13: 145-155, 1983.

11945 - **KOROVIN, A.I., MAMAEV, E.V., MUSIENKO, Z.N., IVANOVA, A.S.:** Urozhaĭnost' yarovogo yachmenya i ovsa pri zamorozhkakh na fone razlichnogo uvlazhneniya pochvy. [Productivity of spring barley and oat under spring frosts and different soil moisture.] - Sel'skokhoz. Biol. 1983(5): 58-60, 1983. [In R, ab: E.]

11946 - **KORTE, L.L., SPECHT, J.E., WILLIAMS, J.H., SORENSEN, R.C.:** Irrigation of soybean genotypes during reproductive ontogeny. II. Yield component responses. - Crop Sci. 23: 528-533, 1983.

11947 - **KORTE, L.L., WILLIAMS, J.H., SPECHT, J.E., SORENSEN, R.C.:** Irrigation of soybean genotypes during reproductive ontogeny. I. Agronomic responses. - Crop Sci. 23: 521-527, 1983.

11948 - **KOWALIK, P.J., TURNER, N.C.:** Diurnal changes in the water relations and transpiration of a soybean crop simulated during the development of water deficits. - Irrig. Sci. 4: 225-238, 1983.

11949 - **KOZINKA, V.:** Metóda pre štúdium rýchlych zmien vodného potenciálu v koreňovom prostredí rastlín. [Method for determination of rapid changes in water potential of substrate around roots.] - In: Dny Rostlinné Fyziologie III. Pp. 217-218. ČSBS, Praha 1983. [In Slov.]

11950 - **KOZLOWSKI, T.T.:** Reduction in yield of forest and fruit trees by water and temperature stress. - In: RAPER, C.D., Jr., KRAMER, P.J. (ed.): Crop Reactions to Water and Temperature Stresses in Humid, Temperate Climates. Pp. 67-88. Westview Press, Boulder 1983.

11951 - **KRAMER, D.:** Genetically determined adaptations in roots to nutritional stress: correlation of structure and function. - In: SARIĆ, M.R., LOUGHMAN, B.C. (ed.): Genetic Aspects of Plant Nutrition. Pp. 33-39. Martinus Nijhoff Publishers, The Hague - Boston - Lancaster 1983.

11952 - **KRAMER, D.:** The possible role of transfer cells in the adaptation of plants to salinity. - Physiol. Plant. 58: 549-555, 1983.

11953 - **KRAMER, P.J.:** Water Relations of Plants. - Academic Press, New York - London - Paris - San Diego - San Francisco - São Paulo - Sydney - Tokyo - Toronto 1983.

11954 - **KRAMER, P.J.:** Problems in water relations of plants and cells. - Int. Rev. Cytol. 85: 253-286, 1983.

11955 - **KRAMER, P.J.:** Water stress research: progress and problems. In: RANDALL, D.D., BLEVINS, D.G., LARSON, R.L., RAPP, B.J., PALLARDY, S.G., NELSON, C.J., POLACCO, J.C., GEORGE, M.F. (ed.): Current Topics in Plant Biochemistry and Physiology. Volume 2. Pp. 129-144. University of Missouri, Columbia 1983.

11956 - **KRAMER, P.J., RAPER, C.D., Jr.:** Research needs and priorities: epilogue. - In: RAPER, C.D., Jr., KRAMER, P.J. (ed.): Crop Reactions to Water and Temperature Stresses in Humid, Temperate Climates. Pp. 359-362. Westview Press, Boulder 1983.

11957 - **KRAUSE, C.R.:** Detection of particulate air pollution injury to red maple leaves with energy dispersive X-ray analysis. - Scanning Electron Microscopy 1983(III): 1493-1497, 1983.

11958 - **KRAUSE, C.R., HOUSTON, D.B.:** Morphological variation in epicuticular wax of SO_2-sensitive and -tolerant eastern white pine clones. - Phytopathology 73: 1266-1269, 1983.

*11959 - **KREUTZER, K., STREBEL, O., RENGER, M.:** Field measurement of seepage and evapotranspiration rate for a soil under plant cover: a comparison of soil water balance and tritium labeling procedure. - J. Hydrol. 48: 137-146, 1980.

11960 - **KRIEDEMANN, P.E., BARRS, H.D.:** Photosynthetic adaptation to water stress and implications for drought resistance. - In: RAPER, C.D., Jr., KRAMER, P.J. (ed.): Crop Reactions to Water and Temperature Stresses in Humid, Temperate Climates. - Pp. 201-230. Westview Press, Boulder 1983.

11961 - **KRIEDEMANN, P.E., SANDS, R., FOSTER, R.C.:** Salt tolerance and root-zone aeration in *Helianthus annuus* (L.) - Growth and stomatal response to solute uptake. - In: MARCELLE, R., CLIJSTERS, H., VAN POUCKE, M. (ed.): Effects of Stress on Photosynthesis. Pp. 313-324. Martinus Nijhoff / Dr W. Junk Publishers, The Hague - Boston - London 1983.

*11962 - **KRIEG, D.R.:** Sorghum. - In: TEARE, I.D., PEET, M.M. (ed.): Crop-Water Relations. Pp. 351-388. John Wiley & Sons, New York - Chichester - Brisbane - Toronto - Singapore 1982.

11963 - **KRIEG, D.R.:** Photosynthetic activity during stress. - Agr. Water Manage. 7: 249-263, 1983. Also in: STONE, J.F., WILLIS, W.O. (ed.): Plant Production and Management under Drought Conditions. Pp. 249-263. Elsevier Scientific Publishing Company, Amstredam 1983.

11964 - **KRIEG, D.R.:** Whole-plant responses to water deficits: carbon assimilation and utilization. - In: TAYLOR, H.M., JORDAN, W.R., SINCLAIR, T.R. (ed.): Limitations to Efficient Water Use in Crop Production. Pp. 319-330. Amer. Soc. Agron., Inc., Crop Sci. Soc. Amer., Inc., Soil Sci. Soc. Amer., Inc., Madison 1983.

*11965 - **KRISHNAN, A., SINGH, S.D., SASTRI, A.S.R.A.S.:** Effect of plant population and nitrogen on water-use efficiency of pearlmillet in the arid zone of western Rajasthan. - Ind. J. agr. Sci. 51: 865-870, 1981.

11966 - **KUBICA, Š.:** Changes in DNA content of differentiating metaxylem of the barley root under conditions of water deficiency. - Biológia (Bratislava) 38: 1057-1062, 1983.

11967 - **KUBIK, M., ANTOSZEWSKI, R.:** Altering the sensitivity of strawberry plants to exogenous abscisic acid. - Physiol. Plant. 57: 505-508, 1983.

11968 - **KUDRYAVETS, R.P.:** Resursy solnechnoĭ ėnergii, tepla i vlagi v usloviyakh Podmoskov'ya i ikh ispol'zovanie v produktsionnom protsesse yabloni. [Resources of solar energy, heath and moisture in Moscow districts and their use in yield formation of apple.] - Sel'skokhoz. Biol. 1983(4): 51-56, 1983. [In R, ab: E.]

11969 - **KUMAR, D., YADAV, J.S.P.:** Salt tolerance of mutants from wheat variety 'HD 1553'. - Ind. J. agr. Sci. 53: 1009-1015, 1983.

11970 - **KUMAR, N., MOHANTY, P.:** Effect of water stress on the photosynthetic activity of greening wheat seedlings. - Biochem. Physiol. Pflanz. 178: 409-412, 1983.

11971 - **KUPKA, J., FOJTÍK, L., NOVÁK, V., GRÉC, L., VOTRUBA, M.:** Vliv meteorologických faktorů na příjem živin a tvorbu sušiny u bobu obecného. [Effect of meteorological factors on mineral nutrition and biomass production in *Vicia faba*.] - In: Dny Rostlinné Fyziologie III. Pp. 223-224. ČSBS, Praha 1983. [In Czech.]

11972 - **KUSHNIRENKO, M.D., OLEKSICH, V.N., KURCHATOVA, G.P., SHTEFYRTSE, A.A., BEĬKAL, M.N.:** Fiziologicheskie osobennosti yabloni pri kapel'nom oroshenii. [Physiological peculiarities of apple-tree under drip irrigation.] - Fiziol. Biokhim. kul't. Rast. 15: 383-388, 1983. [In R, ab: E.]

11973 - **LACHNO, D.R.:** Abscisic acid and indole-3-acetic acid in maize roots subjected to water, salt and mechanical stress. - In: JACKSON, M.B., STEAD, A.D. (ed.): Growth Regulators in Root Development. Pp. 37-54. British Plant Growth Regulator Group, Wantage 1983.

11974 - **LADYMAN, J.A.R., DITZ, K.M., GRUMET, R., HANSON, A.D.:** Genotypic variation for glycinebetaine accumulation by cultivated and wild barley in relation to water stress. - Crop Sci. 23: 465-468, 1983.

11975 - **LAISK, A.:** Calculation of leaf photosynthetic parameters considering the statistical distribution of stomatal apertures. - J. exp. Bot. 34: 1627-1635, 1983.

11976 - **LAKSO, A.N.:** Morphological and physiological adaptations for maintaining photosynthesis under water stress in apple trees. - In: MARCELLE, R., CLIJSTERS, H., VAN POUCKE, M. (ed.): Effects of Stress on Photosynthesis. Pp. 85-93. Martinus Nijhoff / Dr W. Junk Publishers, The Hague - Boston - London 1983.

*11977 - **LAL, J.P., RICHHARIA, A.K.:** Moisture content and radiosensitivity in rice. - Ind. J. Genet. Plant Breed. 41: 309-315, 1981.

11978 - **LANDIVAR, J.A., BAKER, D.N., JENKINS, J.N.:** Application of GOSSYM to genetic feasibility studies. I. Analyses of fruit abscission and yield in okra-leaf cottons. - Crop Sci. 23: 497-504, 1983.

11979 - **LANDIVAR, J.A., BAKER, D.N., JENKINS, J.N.:** Application of GOSSYM to genetic feasibility studies. II. Analyses of increasing photosynthesis, specific leaf weight and longevity of leaves in cotton. - Crop Sci. 23: 504-510, 1983.

11980 - **LANDSBERG, J., WYLIE, F.R.:** Water stress, leaf nutrients and defoliation: a model of dieback of rural eucalypts. - Aust. J. Ecol. 8: 27-41, 1983.

11981 - **LANG, A.:** Turgor-regulated translocation. - Plant Cell Environ. 6: 683-689, 1983.

11982 - **LANGE, O.L., NOBEL, P.S., OSMOND, C.B., ZIEGLER, H.:** Physiological Plant Ecology III. Responses to the Chemical and Biological Environment. (Encyclopedia of Plant Physiology, New Series. Volume 12 C). - Springer-Verlag, Berlin - Heidelberg - New York 1983.

11983 - **LANGE, O.L., NOBEL, P.S., OSMOND, C.B., ZIEGLER, H. (ed.):** Physiological Plant Ecology IV. Ecosystem Processes: Mineral Cycling, Productivity and Man's Influence. (Encyclopedia of Plant Physiology, New Series. Volume 12 D). - Springer-Verlag, Berlin - Heidelberg - New York 1983.

11984 - **LANGE, O.L., REDON, J.:** Epiphytische Flechten im Bereich einer chilenischen "Nebeloase" (Fray Jorge). II. Ökophysiologische Charakterisierung von CO_2--Gaswechsel und Wasserhaushalt. - Flora 174: 245-284, 1983.

*11985 - **LAOUAR, S., VIEIRA da SILVA, J.:** Annual variation of photosynthesis of the olive tree under different watering conditions and related chlorophyll accumulation. - In: MARGARIS, N.S., MOONEY, H.A. (ed.): Components of Productivity of Mediterranean-Climate Regions. Basis and Applied Aspects. Pp. 71-75. Dr W. Junk Publishers, The Hague - Boston - London 1981.

11986 - **LARCHER, W.:** Ökophysiologische Konstitutionseigenschaften von Gebirgspflanzen. - Ber. Deut. bot. Ges. 96: 73-85, 1983.

11987 - **LARCHER, W., WAGNER, J.:** Ökologischer Zeigerwert und physiologische Konstitution von *Sempervivum montanum*. - In: Verhandlungen Gesellschaft für Ökologie. Vol. XI. Pp. 253-264. Göttingen 1983.

11988 - **LASSOIE, J.P., DOUGHERTY, P.M., REICH, P.B., HINCKLEY, T.M., METCALF, C.M., DINA, S.J.:** Ecophysiological investigations of understory eastern redcedar in central Missouri. - Ecology 64: 1355-1366, 1983.

*11989 - **LATIFOV, N.L.:** Metodo de calculo para determinar el coeficiente de lluvia aprovechable. [Calculation method for determining available rainfall coefficient.] - Ciec. Tec. Agr. Cañera 2(2): 49-60, 1982. [In Span, ab: E.]

11990 - **LAWLOR, D.W.:** Integration of biochemical processes in the physiology of water stressed plants. - In: MARCELLE, R., CLIJSTERS, H., VAN POUCKE, M. (ed.): Effects of Stress on Photosynthesis. Pp. 35-44. Martinus Nijhoff / Dr W. Junk Publishers, The Hague - Boston - London 1983.

11991 - **LAWLOR, D.W., KHANNA-CHOPRA, R.:** Regulation of photosynthesis during water stress. - In: Sixth International Congress on Photosynthesis. Abstracts. Volume 1. Pp. 393. Brussels 1983.

11992 - **LAWRENCE, W.T., OECHEL, W.C.:** Effects of soil temperature on the carbon exchange of taiga seedlings. II. Photosynthesis, respiration, and conductance. - Can. J. Forest Res. 13: 850-859, 1983.

11993 - **LAZAN, H.B., BARLOW, E.W.R., BRADY, C.J.:** The significance of vascular connection in regulating senescence of the detached flag leaf of wheat. - J. exp. Bot. 34: 726-736, 1983.

*11994 - **LEACH, J.E., PARKINSON, K.J., WOODHEAD, T.:** Photosynthesis, respiration and evaporation of a field-grown potato crop. - Ann. appl. Biol. 101: 377-390, 1982.

*11995 - **LEACH, J.E., WOODHEAD, T., DAY, W.:** Bias in pressure chamber measurements of leaf water potential. - Agr. Meteorol. 27: 257-263, 1982.

11996 - **LEE, C.H., MOTOSUGI, H., SUGIURA, A., TOMANA, T.:** [Water tolerance of apple trees as affected by different rootstock-scion combinations.] - J. Jap. Soc. hort. Sci. 51: 387-394, 1983. [In Jap, ab: E.]

*11997 - **LEE, K.J., GASKINS, M.H.:** Increased root exudation of ^{14}C-compounds by sorghum seedlings inoculated with nitrogen-fixing bacteria. - Plant Soil 69: 391-399, 1982.

11998 - **LEE, Y., SATTER, R.L.:** Effects of age, water stress, and 1-aminocyclopropane-1-carboxylic acid on leaflet movement in *Albizzia julibrissin*. - Plant Physiol. 71: 669-672, 1983.

*11999 - **LEGEL, S.:** Untersuchungen zu Anbau und Ertragsleistung von Alexandrinerklee (*Trifolium alexandrinum*) in Abhängigkeit von Vegetationsstadium, Aufwuchs und Bewässerung in Ägypten. I. Nährstoffgehalt und Futterwert. - Beitr. trop. Landwirt. Veterinärmedizin 20: 389-399, 1982.

*12000 - **LEGG, B.J.:** Aerial environment and crop growth. - In: ROSE, D.A., CHARLES-EDWARDS, D.A.(ed.): Mathematics and Plant Physiology. Pp. 129-149. Academic Press, London - New York - Toronto - Sydney - San Francisco 1981.

12001 - **LEIGH, R.A.:** Methods, progress and potential for the use of isolated vacuoles in studies of solute transport in higher plant cells. - Physiol. Plant. 57: 390-396, 1983.

12002 - **LEIGH, R.A., JOHNSTON, A.E.:** Concentrations of potassium in the dry matter and tissue water of field-grown spring barley and their relationships to grain yield. - J. agr. Sci. 101: 675-685, 1983.

12003 - **LEIGH, R.A., JOHNSTON, A.E.:** The effects of fertilizers and drought on the concentrations of potassium in the dry matter and tissue water of field--grown spring barley. - J. agr. Sci. 101: 741-748, 1983.

12004 - **LEKHAK, H.D., KUMAR, S., SEN, D.N.:** Ecology of Indian desert. IX. An investigation on the behaviour of leaf and stomata in *Capparis decidua* (Forsk.) Edgew. - Flora 174: 475-478, 1983.

12005 - **LEMON, E.:** Canopy gas exchange: an overview. - In: TAYLOR, H.M., JORDAN, W.R., SINCLAIR, T.R. (ed.): Limitations to Efficient Water Use in Crop Production. Pp. 195-202. Amer. Soc. Agron., Inc., Crop Sci. Soc. Amer., Inc., Soil Sci. Soc. Amer., Inc., Madison 1983.

12006 - **LENDZIAN, K.J., SCHÖNHERR, J.:** In-vivo study of cutin synthesis in leaves of *Clivia miniata* Reg. - Planta 158: 70-75, 1983.

12007 - **LENDZIAN, K.J., UNSWORTH, M.H.:** Ecophysiological effects of atmospheric pollutants. - In: LANGE, O.L., NOBEL, P.S., OSMOND, C.B., ZIEGLER, H. (ed.): Physiological Plant Ecology IV. Ecosystem Processes: Mineral Cycling, Productivity and Man's Influence. Pp. 465-502. Springer-Verlag, Berlin - Heidelberg - New York 1983.

12008 - **LENZ, F., KORNKAMHAENG, P., LEVIN, H.-G.:** Plant reaction to heat stress at low oxygen and high CO_2-concentration. - In: MARCELLE, R., CLIJSTERS, H., VAN POUCKE, M. (ed.): Effects of Stress on Photosynthesis. Pp. 219-226. Martinus Nijhoff / Dr W. Junk Publishers, The Hague - Boston - London 1983.

12009 - **LEONG, S.K., ONG, C.K.:** The influence of temperature and soil water deficit on the development and morphology of groundnut (*Arachis hypogaea* L.). - J. exp. Bot. 34: 1551-1561, 1983.

12010 - **LEOPOLD, A.C.:** Volumetric components of seed imbibition. - Plant Physiol. 73: 677-680, 1983.

12011 - **Le RUDULIER, D., BERNARD, T., POCARD, J.-A., GOAS, G.:** Accroissement le l'osmotolérance chez *Rhizobium meliloti* par la glycine bétaïne et la proline bétaïne. - C.R. Acad. Sci. Paris, Sér. III 297: 155-160, 1983.

12012 - **Le RUDULIER, D., BOUILLARD, L.:** Glycine betaine, an osmotic effector in *Klebsiella pneumoniae* and other members of the *Enterobacteriaceae*. - Appl. environ. Microbiol. 46: 152-159, 1983.

12013 - **LETEY, J., JARRELL, W.M., VALORAS, N., BEVERLY, R.:** Fertilizer application and irrigation management of broccoli production and fertilizer use efficiency. - Agron. J. 75: 502-507, 1983.

12014 - **LEUNING, R.:** Transport of gases into leaves. - Plant Cell Environ. 6: 181-194, 1983.

12015 - **LEVITT, J.:** Plasmolysis shape in relation to freeze-hardening of cabbage plants and to the effect of penetrating solutes. - Plant Cell Environ. 6: 465-470, 1983.

12016 - **LEVY, D.:** Water deficit enhancement of proline and α-amino nitrogen accumulation in potato plants and its association with susceptibility to drought. - Physiol. Plant. 57: 169-173, 1983.

*12017 - **LEVY, D., BEN-HERUT, Z., ALBASEL, N., KAISI, F., MANASRA, I.:** Growing onion seeds in an arid region: drought tolerance and the effect of bulb weight, spacing and fertilization. - Scientia Hort. 14: 1-7, 1981.

*12018 - **LEVY, Y.:** Field determination of free proline accumulation and water-stress in lemon trees. - HortScience 15: 302-303, 1980.

12019 - **LEVY, Y.:** Acclimation of citrus to water stress. - Scientia Hort. 20: 267-273, 1983.

12020 - **LEVY, Y., DODD, J., KRIKUN, J.:** Effect of irrigation, water salinity and rootstock on the vertical distribution of vesicular-arbuscular mycorrhiza in citrus roots. - New Phytol. 95: 397-403, 1983.

*12021 - **LEVY, Y., KRIKUN, J.:** Effect of vesicular-arbuscular mycorrhiza on *Citrus jambhiri* water relations. - New Phytol. 85: 25-32, 1980.

12022 - **LEVY, Y., SYVERTSEN, J.P., NEMEC, S.:** Effect of drought stress and vesicular--arbuscular mycorrhiza on citrus transpiration and hydraulic conductivity of roots. - New Phytol. 93: 61-66, 1983.

12023 - **LI, J.-S., WANG, H.-C., WANG, W.-Y., ZHU, Y-F.:** [Effect of drought on the permeability and membrane lipid composition from maize leaves.] - Acta Phytophys. Sin. 9: 223-229, 1983. [In Chin, ab: E.]

12024 - **LIFSHITZ, R., HANCOCK, J.G.:** Saprophytic development of *Pythium ultimum* in soil as a function of water matric potential and temperature. - Phytopathology 73: 257-261, 1983.

12025 - **LIMA FILHO, J.M.P.:** Temperatura foliar de genótipos de feijão sob duas condições de umidade de solo. [Canopy temperature of bean genotypes under two soil humidity conditions.] - Pesq. agropec. Bras. 18: 703-706, 1983. [In Port, ab: E.]

12026 - **LIN, Z.F., EHLERINGER, J.:** Photosynthetic characteristics of *Amaranthus tricolor*, a C_4 tropical leafy vegetable. - Photosynthesis Res. 4: 171-178, 1983.

12027 - **LINDSTROM, O.M., HUNER, N.P.A., CARTER, J.V.:** Differential thermal analysis of the freezing of water in leaves of cold-hardened and nonhardened puma rye. - Bot. Gaz. 144: 234-239, 1983.

12028 - **LIPTAY, A., SCHOPFER, P.:** Effect of water stress, seed coat restraint, and abscisic acid upon different germination capabilities of two tomato lines at low temperature. - Plant Physiol. 73: 935-938, 1983.

*12029 - **LIS, E.K., ANTOSZEWSKI, R.:** Do growth substances regulate the phloem as well as the xylem transport of nutrients to the strawberry receptacle? - Planta 156: 492-495, 1982.

12030 - **LISS, P.S., CRANE, A.J.:** Man-Made Carbon Dioxide and Climatic Change: A Review of Scientific Problems. - Geo Books, Norwich 1983.

12031 - **LITTLEJOHN, R.O., Jr., WILLIAMS, G.J., III:** Diurnal and seasonal variations in activity of crassulacean acid metabolism and plant water status in a northern latitude population of *Opuntia erinacea*. - Oecologia 59: 83-87, 1983.

*12032 - **LOCKERETZ, W. (ed.):** Agriculture as a Producer and Consumer of Energy. (AAAS Selected Symposium 78.) - Westview Press, Boulder 1982.

*12033 - **LOEBLICH, L.A.**: Photosynthesis and pigments influenced by light intensity and salinity in the halophile *Dunaliella salina* (*Chlorophyta*). - J. mar. Biol. Assoc. U.K. 62: 493-508, 1982.

12034 - **LOFFROY, O., HUBAC, C., VIEIRA da SILVA, J.B.**: Effect of temperature on drought resistance and growth of cotton plants. - Physiol. Plant. 59: 297-301, 1083.

12035 - **LONCHAMP, J.-P., BARRALIS, G.**: Effets de faibles potentiels hydriques sur les possibilités de germination des semences d'*Alopecurus myosuroïdes* Huds. et de *Matricaria perforata* Mérat. - Agronomie 3: 435-441, 1983.

12036 - **LONG, S.P.**: C_4 photosynthesis at low temperatures. - Plant Cell Environ. 6: 345-363, 1983.

12037 - **LONG, S.P., EAST, T.M., BAKER, N.R.**: Chilling damage to photosynthesis in young *Zea mays*. I. Effects of light and temperature variation on photosynthetic CO_2 assimilation. - J. exp. Bot. 34: 177-188, 1983.

12038 - **LOOMIS, R.S.**: Crop manipulations for efficient use of water: an overview. - In: TAYLOR, H.M., JORDAN, W.R., SINCLAIR, T.R. (ed.): Limitations to Efficient Water Use in Crop Production. Pp. 345-374. Amer. Soc. Agron., Inc., Crop Sci. Soc. Amer., Inc., Soil Sci. Soc. Amer., Inc., Madison 1983.

*12039 - **LOSAVIO, N., VENEZIAN, M.L., ZERBI, G.**: Sunflower (*Helianthus annuus* L.) response to increasing irrigation levels in southern Italy. - In: IX Conferencia Internacional del Girasol. Tomo II. (Pubbl. N. 52). Pp. 98-109. Consiglio Nazionale delle Ricerche, Napoli-Ponticelli 1981.

12040 - **LU, M.-H., XU, M.-D., SHEN, Z.-D., DING, J.**: [Leaf senescence and stomatal aperture.] - Acta Phytophysiol. Sin. 9: 61-68, 1983. [In Chin, ab: E.]

12041 - **LUARD, E.J.**: Two problems encountered in preparation of fungi grown at low osmotic potential for microscopy. - Trans. Brit. mycol. Soc. 80: 529-533, 1983.

12042 - **LUARD, E.J.**: Activity of isocitrate dehydrogenase from three filamentous fungi in relation to osmotic and solute effects. - Arch. Microbiol. 134: 233-237, 1983.

12043 - **LUDLOW, M.M., CHU, A.C.P., CLEMENTS, R.J., KERSLAKE, R.G.**: Adaptation of species of *Centrosema* to water stress. - Aust. J. Plant Physiol. 10: 119-130, 1983.

*12044 - **LUK'YANOVA, N.M., RADCHENKO, S.S.**: O tsirkadnoĭ ritmike vodnogo obmena vechnozelenykh rasteniĭ v usloviyakh kruglosutochnogo iskusstvennogo osveshcheniya. [Circadian rhytmics of water exchange in ever-green plants under twenty-four-hour artificial lighting conditions.] - Byull. gosud. Nikitskogo bot. Sada 47: 77-81, 1982. [In R, ab: E.]

*12045 - **LUQUE, A.A., BINGHAM, F.T.**: The effect of the osmotic potential and specific ion concentration of the nutrient solution on the uptake and reduction of nitrate by barley seedlings. - Plant Soil 63: 227-237, 1981.

*12046 - **LUVALL, J.C., MURPHY, C.E., Jr.**: Evaluation of the tritiated water method for measurement of transpiration in young *Pinus taeda* L. - Forest Sci. 28: 5-16, 1982.

12047 - **LUXMOORE, R.J.**: Water budget of an eastern deciduous forest stand. - Soil Sci. Soc. Amer. J. 47: 785-791, 1983.

12048 - **MA, P.C., HUNT, L.A.:** Effects of two photoperiod treatments on leaf photosynthesis throughout ontogeny in 'Fergus' barley. - Can. J. Bot. 61: 792-797, 1983.

12049 - **MAAS, E.V., HOFFMAN, G.J., CHABA, G.D., POSS, J.A., SHANNON, M.C.:** Salt sensitivity of corn at various growth stages. - Irrig. Sci. 4: 45-57, 1983.

*12050 - **MacFARLANE, J.D., KERSHAW, K.A.:** Physiological-environmental interactions in lichens. IX. Thermal stress and lichen ecology. - New Phytol. 84: 669-685, 1980.

12051 - **MacFARLANE, J.D., KERSHAW, K.A., WEBBER, M.R.:** Physiological-environmental interactions in lichens. XVII. Phenotypic differences in the seasonal pattern of net photosynthesis in *Cladonia rangiferina*. - New Phytol. 94: 217-233, 1983.

12052 - **MACKAY, M.A., NORTON, R.S., BOROWITZKA, L.J.:** Marine blue-green algae have a unique osmoregulatory system. - Mar. Biol. 73: 301-307, 1983.

*12053 - **MACKERRON, D.K.L.:** Growth and water use in the red raspberry (*Rubus idaeus* L.) I. Growth and yield under different levels of soil moisture stress. - J. hort. Sci. 57: 295-306, 1982.

*12054 - **MacROBBIE, E.A.C.:** Osmotic measurements on stomatal cells of *Commelina communis* L. - J. Membrane Biol. 53: 189-198, 1980.

*12055 - **MacROBBIE, E.A.C.:** Chloride transport in stomatal guard cells. - Phil. Trans. roy. Soc. London B 299: 469-481, 1982.

12056 - **MacROBBIE, E.A.C.:** Effects of light/dark on cation fluxes in guard cells of *Commelina communis* L. - J. exp. Bot. 34: 1695-1710, 1983.

*12057 - **MacROBBIE, E.A.C., LETTAU, J.:** Ion content and aperture in "isolated" guard cells of *Commelina communis* L. - J. Membrane Biol. 53: 199-205, 1980.

*12058 - **MacROBBIE, E.A.C., LETTAU, J.:** Potassium content and aperture in "intact" stomatal and epidermal cells of *Commelina communis* L. - J. Membrane Biol. 56: 249-256, 1980.

12059 - **MADHAVAN, S., CHROMINISKI, A., SMITH, B.N.:** Effect of ethylene on stomatal opening in tomato and carnation leaves. - Plant Cell Physiol. 24: 569-572, 1983.

12060 - **MAGNUSSEN, S.:** Untersuchungen über den Einfluss der Luftfeuchtigkeit auf den Transpirationswiderstand junger licht- und schattengeprägter Weiss- und Küstentannen. - Flora 173: 279-291, 1983.

*12061 - **MAHALL, B.E., SCHLESINGER, W.H.:** Effects of irradiance on growth, photosynthesis, and water use efficiency of seedlings of the chaparral shrub, *Ceanothus megacarpus*. - Oecologia 54: 291-299, 1982.

*12062 - **MAHMOUD, E.A., HILL, M.J.:** Salt tolerance of sugar beet at various temperatures. - N. Zeal. J. agr. Res. 24: 67-71, 1981.

12063 - **MAHON, J.D.:** Limitations to the use of physiological variability in plant breeding. - Can. J. Plant Sci. 63: 11-21, 1983.

12064 - **MAIER-MAERCKER, U.:** The role of peristomatal transpiration in the mechanism of stomatal movement. - Plant Cell Environ. 6: 369-380, 1983.

12065 - **MAIER-MAERCKER, U.:** A critical assessment of the role of potassium and osmolarity in stomatal opening. - J. exp. Bot. 34: 811-824, 1983.

12066 - **MAILLOT, C., PENNARUN, A.-M., HELLER, R.:** Ajustement hydrique des cellules libres d'*Acer pseudoplatanus* adaptées à des concentrations élevées de NaCl. - Physiol. vég. 21: 501-507, 1983.

12067 - **MAĬSNER, A.D.:** Ab sposabakh atsėnki i rėgulyatsyi vodnaga rėzhymu pasevaŭ. [Evaluation and control of water relations in stands.] - Vestsi Akad. Navuk Belorus. SSR, Ser. biyal. Navuk 1983(6): 18-21, 1983. [In Belorus, ab: E, R.]

*12068 - **MAJOR, D.J., HAMMAN, W.M.:** Comparison of sorghum with wheat and barley grown on dryland. - Can. J. Plant Sci. 61: 37-43, 1981.

*12069 - **MAKARIEV, Z., DOĬCHEV, K., LEKHOVA, E.:** Fotosintetichen potentsial na mladi yab"lkovi d"rveta, otglezhdani pri razlichni rezhimi na napoyavane i azotno torene. [Photosynthetic potential in young apple trees affected by different irrigation and nitrogen nutrition.] - Gradin. lozar. Nauka 18(7): 10-15, 1981. [In Bulg, ab: E, R.]

12070 - **MALIKOV, V.G., PEREPELYATNIKOV, G.P., ALEKSAKHIN, R.M.:** Nakoplenie radiostrontsiya v rasteniyakh kukuruzy, lyutserny i kapusty iz pochvy i vody v usloviyakh orosheniya. [Accumulation of radiostrontium in maize, alfalfa and cabbage plants from soil and irrigation water.] - Sel'skokhoz. Biol. 1983(3): 76-78, 1983. [In R.]

12071 - **MALIKOV, V.G., PEREPELYATNIKOV, G.P., ALEKSAKHIN, R.M., LISTROVAYA, V.A.:** Nakoplenie ^{89}Sr i ^{134}Cs v ozimoĭ pshenitse pri oroshenii dozhdevaniem. [Accumulation of ^{89}Sr and ^{134}Cs in winter wheat plants under sprinkler irrigation.] - Sel'skokhoz. Biol. 1983(7): 68-71, 1983. [In R, ab: E.]

*12072 - **MALYSHEVA, G.S., GANNIBAL, B.K., IZMAĬLOVA, N.N.:** Reaktsiya travyanistykh rasteniĭ fistashnikov Badkhyza na usloviya sredy. [Reaction of grasses in pistache forest of Badkhyze to environmental conditions.] - Ėkologiya 1982(5): 8-15, 1982. [In R.]

12073 - **MAMBANI, B., LAL, R.:** Response of upland rice varieties to drought stress. I. Relation between root system development and leaf water potential. - Plant Soil 73: 59-72, 1983.

12074 - **MAMBANI, B., LAL, R.:** Response of upland rice varieties to drought stress. II. Screening rice varieties by means of variable moisture regimes along a toposequence. - Plant Soil 73: 73-94, 1983.

12075 - **MAMBANI, B., LAL, R.:** Response of upland rice varieties to drought stress. III. Estimating root system configuration from soil moisture data. - Plant Soil 73: 95-104, 1983.

*12076 - **MANN, R.K., RIECK, C.E., WITT, W.W.:** Germination and emergence of burcucumber (*Sicyos angulatus*). - Weed Sci. 29: 83-86, 1981.

*12077 - **MANNER, R.:** Biomass production: A plant breeder's view. - In: Improvement of Oil-Seed and Industrial Crops by Induced Mutations. Pp. 265-281. International Atomic Energy Agency, Vienna 1982.

12078 - **MANSFIELD, T.A.:** Movements of stomata. - Sci. Prog. (Oxford) 68: 519-542, 1983.

12079 - **MANSFIELD, T.A., PEMADASA, M.A., SNAITH, P.J.:** New possibilities for controlling foliar absorption via stomata. - Pestic. Sci. 14: 294-298, 1983.

*12080 - **MANTELL, A., GOLDSCHMIDT, E.E., MONSELISE, S.P.:** Turnover of tritiated water in calamondin plants. - J. Amer. Soc. hort. Sci. 105: 741-744, 1980.

12081 - **MARCELLE, R., CLIJSTERS, H., VAN POUCKE, M. (ed.):** Effects of Stress on Photosynthesis. - Martinus Nijhoff / Dr W. Junk Publishers, The Hague - Boston - London 1983.

12082 - **MAREK, M., SAILEROVÁ, E.:** The influence of nitrogen nutrition on photosynthetic characteristics and water relation in young barley leaves. - In: Sixth International Congress on Photosynthesis. Abstracts. Volume 2. Pp. 381. Brussels 1983.

12083 - **MARQUES, J.B.B., LIN, S.S.:** Rendimento da soja em diferentes populações, espaçamentos e épocas de irrigação. [Soybean yield in different plant populations, spacings, and irrigation periods.] - Pesq. agropec. Bras. 18: 1089-- 1097, 1983. [In Port, ab: E.]

*12084 - **MARTIN, C.E., CHRISTENSEN, N.L., STRAIN, B.R.:** Seasonal patterns of growth, tissue acid fluctuations, and $^{14}CO_2$ uptake in the Crassulacean Acid Metabolism epiphyte *Tillandsia usneoides* L. (Spanish moss). - Oecologia 49: 322-328, 1981.

12085 - **MARTIN, C.E., ZEE, A.K.:** C_3 photosynthesis and Crassulacean acid metabolism in a Kansas rock outcrop succulent, *Talinum calycinum* Engelm. (*Portulacaceae*). - Plant Physiol. 73: 718-723, 1983.

12086 - **MARTIN, E.S., DONKIN, M.E., STEVENS, R.A.:** Stomata. (Studies in Biology No. 155). - Edward Arnold, London 1983.

12087 - **MARTIN, F.W.:** Variation of sweet potatoes with respect to the effects of waterlogging. - Trop. Agr. 60: 117-121, 1983.

12088 - **MARTIN, M.W., MILLER, D.E.:** Variations in responses of potato germplasm to deficit irrigation as affected by soil texture. - Amer. Potato J. 60: 671-684, 1983.

12089 - **MARTIN, R.J., CARNAHAN, J.A.:** The effect of field storage and laboratory conditions on germination of five *Xanthium* species. - Aust. J. agr. Res. 34: 249-260, 1983.

*12090 - **MARTIN, S.S.:** Sugarbeet. - In: TEARE, I.D., PEET, M.M. (ed.): Crop-Water Relations. Pp. 423-444. John Wiley & Sons, New York - Chichester - Brisbane - Toronto - Singapore 1982.

12091 - **MASON, W.K., MEYER, W.S., BARRS, H.D., SMITH, R.C.G.:** A comparison of irrigated maize growth in southern New South Wales on sites with and without a preceding fallow. - Aust. J. exp. Agr. anim. Husb. 23: 58-63, 1983.

12092 - **MASON, W.K., MEYER, W.S., SMITH, R.C.G., BARRS, H.D.:** Water balance of three irrigated crops on fine-textured soils of the Riverine Plain. - Aust. J. agr. Res. 34: 183-191, 1983.

12093 - **MASSMAN, W.J.:** The derivation and validation of a new model for the interception of rainfall by forests. - Agr. Meteorol. 28: 261-286, 1983.

*12094 - **MASUDA, M., GOMI, K.:** Diurnal changes of the exudation rate and the mineral concentration in xylem sap after decapitation of grafted and non-grafted cucumbers. - J. Jap. Soc. hort. Sci. 51: 293-298, 1982.

12095 - **MATHENY, T.A., HUNT, P.G.:** Effects of irrigation on accumulation of soil and symbiotically fixed N by soybean grown on a Norfolk loamy sand. - Agron. J. 75: 719-722, 1983.

*12096 - **MATHEW, L., SHAH, G.L.:** Structure and ontogeny of stomata on the vegetative and floral organs in nine species of *Verbena*. - Proc. Ind. Acad. Sci. - Plant. Sci. 90: 485-497, 1981.

12097 - **MATHUR, D.D., BHAGSARI, A.S.:** Effect of photosynthetically active radiation, temperature, and antitranspirants on photosynthesis and respiration of letherleaf fern. - HortScience 18: 189-191, 1983.

*12098 - **MATSUMOTO, Y., SUKIGARA, N., KAKUBARI, Y., NEGISI, K.:** [A system for measuring photosynthesis and respiration rates of standing trees and its application to a *Fagus crenata* stand.] - J. Jap. Forest Soc. 64: 149-154, 1982. [In Jap.]

*12099 - **MATSUSHIMA, J., YONEMORI, K.:** Effect of ethylene on fruit drop, growth, photosynthesis, diffusive resistance and abscisic acid of citrus trees. - Proc. int. Soc. Citricult. 1: 304-307, 1981.

12100 - **MATTEONI, J.A., SINCLAIR, W.A.:** Stomatal closure in plants infected with mycoplasmalike organisms. - Phytopathology 73: 398-402, 1983.

12101 - **MATTHES, U., FEIGE, G.B.:** Ecophysiology of lichen symbioses. - In: LANGE, O.L., NOBEL, P.S., OSMOND, C.B., ZIEGLER, H. (ed.): Physiological Plant Ecology III. Responses to the Chemical and Biological Environment. Pp. 423-467. Springer-Verlag, Berlin - Heidelberg - New York 1983.

12102 - **MATTHEWS, M.A., BOYER, J.S.:** Acclimation of photosynthesis to low leaf water potentials in sunflower. - In: Sixth International Congress on Photosynthesis. Abstracts. Volume 1. Pp. 395. Brussels 1983.

*12103 - **MATUSZKIEWICZ, J.M.:** Ground water as the main factor differentiating the natural vegetation in a landscape phytocomplex "Grabowy" range in the Kampinos Forest Complex. - Acta Soc. Bot. Pol. 51: 309-321, 1982.

12104 - **MAUK, C.S., BREEN, P.J., MACK, H.J.:** Yield response of major pod-bearing nodes in bush snap beans to irrigation and plant population. - J. Amer. Soc. hort. Sci. 108: 935-939, 1983.

*12105 - **MAURYA, P.R., LAL, R.:** Effect of water table depth and tillage methods on plant-water status and yield of rice. - Plant Soil 59: 17-22, 1981.

12106 - **MAWSON, B.T., COLMAN, B.:** The inhibition of photosynthesis and photorespiration in isolated mesophyll cells of *Phaseolus* and *Lycopersicon* by reduced osmotic potentials. - Physiol. Plant. 57: 21-27, 1983.

12107 - **MAXWELL, E.L.:** Remote monitoring of rangeland production. - Agr. Water Manage. 7: 323-340, 1983. Also in: STONE, J.F., WILLIS, W.O. (ed.): Plant Production and Management under Drought Conditions. Pp. 323-340. Elsevier Scientific Publishing Company, Amsterdam 1983.

12108 - **McANENEY, K.J., JUDD, M.J.:** Pasture production and water use measurements in the central Waikato. - N. Zeal. J. agr. Res. 26: 7-13, 1983.

12109 - **McCONATHY, R.K.:** Tulip-poplar leaf diffusion resistance calculated from stomatal dimensions and varying environmental parameters. - Forest Sci. 29: 139-148, 1983.

12110 - **McDONNELL, E.M., COUGHLAN, S.J., WYN JONES, R.G.:** Differential effects of abscisic acid on glycinebetaine and proline accumulation in three plant species. - Z. Pflanzenphysiol. 109: 207-213, 1983.

12111 - **McINTYRE, G.I.:** Evidence of transpiration-induced curvatures in the rhizome of *Agropyron repens* L. Beauv. - Ann. Bot. 51: 395-398, 1983.

12112 - **McINTYRE, G.I.:** The role of water distribution in plant tropisms. - Aust. J. Plant Physiol. 7: 401-413, 1983.

12113 - **McINTYRE, G.I., HSAIO, A.I.:** The role of water in the mechanism of seed dormancy in wild oats (*Avena fatua* L.). - In: SMITH, A.E. (ed.): Wild Oat Symposium Proceedings. No. 12. Pp. 187-200. Canadian Plains Research Center, Regina 1983.

*12114 - **McINTYRE, G.I., HSIAO, A.I.:** Influence of nitrogen and humidity on rhizome bud growth and glyphosate translocation in quackgrass (*Agropyron repens*). - Weed Sci. 30: 655-660, 1982.

12115 - **McKENZIE, R.C., SPROUT, C.H., CLARK, N.F.:** The relationship of the yield of irrigated barley to soil salinity as measured by several methods. - Can. J. Soil Sci. 63: 519-528, 1983.

*12116 - **McLAUGHLIN, S.B., TAYLOR, G.E.:** Relative humidity: Important modifier of pollutant uptake by plants. - Science 211: 167-168, 1981.

12117 - **McLENDON, B.D., THOMSON, S.J., CHESNESS, J.L.:** Irrigation scheduling - a valid option with microprocessor-based controls. - Agr. Eng. 64: 12-14, 1983.

12118 - **McMURTRIE, R., WOLF, L.:** A model of competition between trees and grass for radiation, water and nutrients. - Ann. Bot. 52: 449-458, 1983.

12119 - **McNAUGHTON, K.G.:** The direct effect of shelter on evaporation rates: theory and an experimental test. - Agr. Meteorol. 29: 125-136, 1983.

12120 - **McNAUGHTON, K.G., JARVIS, P.G.:** Predicting effects of vegetation changes on transpiration and evaporation. - In: KOZLOWSKI, T.T. (ed.): Water Deficits and Plant Growth. Volume VII. Additional Woody Crop Plants. Pp. 2-47. Academic Press, New York - London - Paris - San Diego - San Francisco - São Paulo - Sydney - Tokyo - Toronto 1983.

12121 - **McNAUGHTON, S.J., WALLACE, L.L., COUGHENOUR, M.B.:** Plant adaptation in an ecosystem context: effects of defoliation, nitrogen, and water on growth of an african C_4 sedge. - Ecology 64: 307-318, 1983.

12122 - **McWILLIAM, J.R.:** Physiological basis for chilling stress and the consequences for crop production. - In: RAPER, C.D., Jr., KRAMER, P.J. (ed.): Crop Reactions to Water and Temperature Stresses in Humid, Temperate Climates. Pp. 113-132. Westview Press, Boulder 1983.

12123 - **MEDERSKI, H.J.:** Effects of water and temperature stress on soybean plant growth and yield in humid, temperate climates. - In: RAPER, C.D., Jr., KRAMER, P.J. (ed.): Crop Reactions to Water and Temperature Stresses in Humid, Temperate Climates. Pp. 35-48. Westview Press, Boulder 1983.

12124 - **MEI, H.-S., YANG, J.-J.:** [A comparative study of inhibiting stomatal opening between the humate and phytohormones.] - Acta Phytophysiol. Sin. 9: 143-149, 1983. [In Chin, ab: E.]

12125 - **MEIDNER, H.:** Our understanding of plant water relations. - J. exp. Bot. 34: 1606-1618, 1983.

12126 - **MEINZER, F., SEYMOUR, V., GOLDSTEIN, G.:** Water balance in developing leaves of four tropical savanna woody species. - Oecologia 60: 237-243, 1983.

*12127 - **MEIRI, A., PLAUT, Z., PINCAS, L.:** Salt tolerance of glasshouse-grown muskmelon. - Soil Sci. 131: 189-193, 1981.

12128 - **MESHCHERYAKOV, A.B., KHOLODOVA, V.P., EHWALD, R.:** Hexose uptake in *Beta* root parenchyma stimulation by wounding, washing and reduced water potential, inhibition by different electrolytes. - Biochem. Physiol. Pflanz. 178: 273-277, 1983.

12129 - **MÉTRAUX, J.-P., KENDE, H.:** The role of ethylene in the growth response of submerged deep water rice. - Plant Physiol. 72: 441-446, 1983.

*12130 - **MEUSEL, F.:** Selektive Identifikation - ein Verfahren zur Bestimmung von Parametern für ein Entzugsmodell der Beregnungsteuerung. - Arch. Acker-Pflanzenbau Bodenk. 26: 749-754, 1982.

*12131 - **MEYERS, S.P., NICHOLS, S.L., BAER, G.R., MOLIN, W.T., SCHRADER, L.E.:** Ploidy effects in isogenic populations of alfalfa I. Ribulose-1,5-bisphosphate carboxylase, soluble protein, chlorophyll, and DNA in leaves. - Plant Physiol. 70: 1704-1709, 1982.

*12132 - **MICHAEL, G., FEILER, S., RANFT, H., TESCHE, M.:** Der Einfluss von Schwefeldioxid und Frost auf Fichten (*Picea abies* (L.) Karst.). - Flora 172: 317-326, 1982.

12133 - **MICHAUX-FERRIERE, N., MANSOUR, K., HALLET, J.-N.:** Stress hydrique et arrêt du cycle cellulaire chez les bryophytes et les ptéridophytes. - Ann. Sci. Natur. Bot. (Paris) 5: 195-209, 1983.

12134 - **MICHEL, B.E.:** Evaluation of the water potentials of solutions of polyethylene glycol 8000 both in the absence and presence of other solutes. - Plant Physiol. 72: 66-70, 1983.

12135 - **MICHEL, B.E., WIGGINS, O.K., OUTLAW, W.H., Jr.:** A guide to establishing water potential of aqueous two-phase solutions (polyethylene glycol plus dextran) by amendement with mannitol. - Plant Physiol. 72: 60-65, 1983.

12136 - **MILBORROW, B.V.:** Pathways to and from abscisic acid. - In: ADDICOTT, F.T. (ed.): Abscisic Acid. Pp. 79-111. Praeger Publishers, New York 1983.

12137 - **MILLER, D.E., BURKE, D.W.:** Response of dry beans to daily deficit sprinkler irrigation. - Agron. J. 75: 775-778, 1983.

12138 - **MILLER, D.E., MARTIN, M.W.:** Effect of daily irrigation rate and soil texture on yield and quality of Russet Burbank potatoes. - Amer. Potato J. 60: 745-758, 1983.

*12139 - **MILLER, M.N., GUNTER, D.L., MELTON, B.E., JOHNSON, C.R., MIDCAP, J.T.:** Selecting economically optimal levels of fertilization and irrigation for container production of woody ornamentals. - J. Amer. Soc. hort. Sci. 105: 766-768, 1980.

12140 - **MILLER, P.C.:** Canopy structure of mediterranean-type shrubs in relation to heat and moisture. - In: KRUGER, F.J., MITCHELL, D.T., JARVIS, J.U.M. (ed.): Mediterranean-Type Ecosystems. Pp. 133-160. Springer-Verlag, Berlin - Heidelberg - New York 1983.

12141 - **MILLER, P.C.:** Comparison of water balance characteristics of plant species in "natural" versus modified ecosystems. - In: MOONEY, H.A., GORDON, M. (ed.): Disturbance and Ecosystems. Components of Response. Pp. 188-212. Springer-Verlag, Berlin - Heidelberg - New York 1983.

12142 - **MILLER, P.C., MILLER, J.M., MILLER, P.M.:** Seasonal progression of plant water relations in Fynbos in the western Cape Province, South-Africa. - Oecologia 56: 392-396, 1983.

12143 - **MILLER, P.C., POOLE, D.K., MILLER, P.M.:** The influence of annual precipitation, topography, and vegetative cover on soil moisture and summer drought in southern California. - Oecologia 56: 385-391, 1983.

12144 - **MILLETTE, J.A.:** Effect of water table depths on the growth of carrots and onions on an organic soil in vitro. - Can. J. Plant Sci. 63: 739-746, 1983.

12145 - **MINTER, T.C., LORD, E.M.:** Effects of water stress, abscisic acid, and gibberellic acid on flower production and differentiation in the cleistogamous species *Collomia grandiflora* Dougl. ex Lindl. (*Polemoniaceae*). - Amer. J. Bot. 70: 618-624, 1983.

*12146 - **MISRA, G., SINGH, K.P.:** Effect of soil moisture and clipping stresses on the nutrient (N, P and K) concentration, uptake and use efficiency in one temperate and two tropical grasses. - Plant Soil 69: 413-421, 1982.

12147 - **MITAL, D., SUCOFF, E.:** Predicting soil moisture depletion beneath trembling aspen. - Can. J. Forest Res. 13: 45-52, 1983.

12148 - **MITCHELL, P.D., CHALMERS, D.J.:** A comparison of microjet and point emitter (trickle) irrigation in the establishment of a high-density peach orchard. - HortScience 18: 472-474, 1983.

12149 - **MIYAMOTO, S.:** Consumptive water use of irrigated pecans. - J. Amer. Soc. hort. Sci. 108: 676-681, 1983.

*12150 - **MIZRAHI, Y.:** Effect of salinity on tomato fruit ripening. - Plant Physiol. 69: 966-970, 1982.

*12151 - **MOKHAMED, T.S.:** V"rkhu evapotranspiratsiyata na lyutsernata i neĭnoto prognozirane. [Alfalfa evapotranspiration and its prediction.] - Rasteniev. Nauki 19(8): 64-70, 1982. [In Bulg, ab: E, R.]

12152 - **MOLDAU, H., RAHI, M.:** Enhancement of maintenance respiration under water stress. - In: MARCELLE, R., CLIJSTERS, H., VAN POUCKE, M. (ed.): Effects of Stress on Photosynthesis. Pp. 121-131. Martinus Nijhoff / Dr W. Junk Publishers, The Hague - Boston - London 1983.

*12153 - **MOLNAR, I., BELIĈ, B., PEKANOVIĈ, V.:** Uticaj vlažnosti zemljišta i vazduha na prinos i kvalitet semena soje. [Effect of soil moisture and air humidity on yield formation and grain quality in some soybean varieties.] - Arhiv poljoprivred. Nauke 43: 445-454, 1982. [In Serb, ab: E.]

12154 - **MOLOTKOVSKIĬ, Yu.G., YAKOVLEVA, G.A.:** Mekhanizm ėnergozavisimogo pogloshcheniya ionov izolirovannymi khloroplastami. [Mechanism of energy-dependent ion uptake by isolated chloroplasts.] - Fiziol. Rast. 30: 467-476, 1983. [In R, ab: E.]

12155 - **MOLYNEUX, D.E., DAVIES, W.J.:** Rooting pattern and water relations of three pasture grasses growing in drying soil. - Oecologia 58: 220-224, 1983.

*12156 - **MONK, L.S., BRÄNDLE, R.:** Adaptation of respiration and fermentation to changing levels of oxygen in rhizomes of *Schoenoplectus lacustris* (L.) Palla and its significance to flooding tolerance. - Z. Pflanzenphysiol. 105: 369-374, 1982.

*12157 - **MONNIER, G., CABIBEL, B., GUENNELON, R., HABIB, R.:** Problèmes posés par l'alimentation hydrique et minérale des arbres fruitiers sous irrigation localisée. - Fruits 37: 189-193, 1982.

12158 - **MONSON, R.K., LITTLEJOHN, R.O., Jr., WILLIAMS, G.J., III:** Photosynthetic adaptation to temperature in four species from the Colorado shortgrass steppe: a physiological model for coexistence. - Oecologia 58: 43-51, 1983.

*12159 - **MONSON, R.K., STIDHAM, M.A., WILLIAMS, G.J., III, EDWARDS, G.E., URIBE, E.G.:** Temperature dependence of photosynthesis in *Agropyron smithii* Rydb. I. Factors affecting net CO_2 uptake in intact leaves and contribution from ribulose-1,5-bisphosphate carboxylase measured *in vivo* and *in vitro*. - Plant Physiol. 69: 921-928, 1982.

12160 - **MOONEY, H.A.:** Carbon-gaining capacity and allocation patterns of mediterranean-climate plants. - In: KRUGER, F.J., MITCHEL, D.T., JARVIS, J.U.M. (ed.): Mediterranean-Type Ecosystems. The Role of Nutrients. Pp. 103-119. Springer--Verlag, Berlin - Heidelberg - New York 1983.

12161 - **MOONEY, H.A., CHU, C.:** Stomatal responses to humidity of coastal and interior populations of a Californian shrub. - Oecologia 57: 148-150, 1983.

12162 - **MOONEY, H.A., FIELD, C., WILLIAMS, W.E., BERRY, J.A., BJÖRKMAN, O.:** Photosynthetic characteristics of plants of a Californian cool coastal environment. - Oecologia 57: 38-42, 1983.

12163 - **MORESHET, S., COHEN, Y., FUCHS, M.:** Response of mature 'Shamouti' orange trees to irrigation of different soil volumes at similar levels of available water. - Irrig. Sci. 3: 223-236, 1983.

12164 - **MORGAN, D.C., ROOK, D.A., WARRINGTON, I.J., TURNBULL, H.L.:** Growth and developement of *Pinus radiata* D. Don: the effect of light quality. - Plant Cell Environ. 6: 691-701, 1983.

12165 - **MORGAN, J.A., WILLIS, W.O.:** Gas exchange and water relations of 'Olaf' spring wheat. - Crop Sci. 23: 541-546, 1983.

12166 - **MORISON, J.I.L., GIFFORD, R.M.:** Stomatal sensitivity to carbon dioxide and humidity. A comparison of two C_3 and two C_4 grass species. - Plant Physiol. 71: 789-796, 1983.

12167 - **MORISON, J.I.L., JARVIS, P.G.:** Direct and indirect effects of light on stomata. I. In Scots pine and Sitka spruce. - Plant Cell Environ. 6: 95-101, 1983.

12168 - **MORISON, J.I.L., JARVIS, P.G.:** Direct and indirect effects of light on stomata. II. In *Commelina communis* L. - Plant Cell Environ. 6: 103-109, 1983.

*12169 - **MORRIS, J.R., CAWTHON, D.L.:** Effect of irrigation, fruit load, and potassium fertilization on yield, quality, and petiole analysis of Concord (*Vitis labrusca* L.) grapes. - Amer. J. Enol. Viticult. 33: 145-148, 1982.

12170 - **MORRIS, J.R., SPAYD, S.E., CAWTHON, D.L.:** Influence of drip irrigation on 'Concord' grapes. - Arkansas Farm Res. 32: 5, 1983.

12171 - **MORTON, F.I.:** Operational estimates of areal evapotranspiration and their significance to the science and practice of hydrology. - J. Hydrol. 66: 1-76, 1983.

*12172 - **MORTON, J.B., MATHRE, D.E.:** Physiological effects of *Cephalosporium gramineum* on growth and yield of winter wheat cultivars. - Phytopathology 70: 807-811, 1980.

12173 - **MOSER, T.J., NASH, T.H., III, LINK, S.O.:** Diurnal gross photosynthetic patterns and potential seasonal CO_2 assimilation in *Cladonia stellaris* and *Cladonia rangiferina*. - Can. J. Bot. 61: 642-655, 1983.

*12174 - **MOSTAGHIMI, S., MITCHELL, J.K.:** Trickle irrigation of vegetable crops in the subhumid midwest. - In: Proceedings of the Specialty Conference on Environmentally Sound Water and Soil Management. Pp. 484-493. ASCE, Orlando 1982.

12175 - **MOUGOU, A., LEMEUR, R., SCHALCK, J.:** Effects of increasing water stress on the functional photosynthetic characteristics of several tomato species and one hybrid. - In: Sixth International Congress on Photosynthesis. Abstracts. Volume 1. Pp. 396. Brussels 1983.

12176 - **MUKHERJEE, S.P., CHOUDHURI, M.A.:** Effect of some reducing agents on water stress-induced oxidative and deteriorative processes of *Vigna* seedlings. - Biol. Plant. 25: 401-407, 1983.

12177 - **MUKHERJEE, S.P., CHOUDHURI, M.A.:** Imlications of water stress-induced changes in the levels of endogenous ascorbic acid and hydrogen peroxide in *Vigna* seedlings. - Physiol. Plant. 58: 166-170, 1983.

12178 - **MUKHINA, V.A., SELIVANOVA-GORODKOVA, E.A.:** Okislitel'nye fermenty i vodouderzhivayushchaya sposobnost' list'ev *Podophyllum peltatum* i *P. hexandrum* S.L. (*Podophyllaceae*). [Oxidizing enzymes and waterholding ability of the leaves of *Podophyllum peltatum* and *P. hexandrum* S.L. (*Podophyllaceae*).] - Bot. Zh. 68: 62-69, 1983. [In R.]

*12179 - **MUKHIYA, Y.K., SHROTRIYA, N., JOSHI, J.K., SINGH, V.P.:** Salt tolerance of wheat, barley and soybean in respect of germination and pigment concentration. - Ind. J. agr. Sci. 51: 881-885, 1981.

*12180 - **MUNDEL, G.:** Untersuchungen über die Evapotranspiration von Grasland auf Grundwasserstandorten. 1. Mitteilung: Beziehungen zwischen meteorologischen Faktoren und Evapotranspiration. - Arch. Acker- Pflanzenbau Bodenk. 26: 507-513, 1982.

*12181 - **MUNDEL, G.:** Untersuchungen über die Evapotranspiration von Grasland auf Grundwasserstandorten. 2. Mitteilung: Beziehungen zwischen Bodenfaktoren und Evapotranspiration. - Arch. Acker- Pflanzenbau Bodenk. 26: 515-521, 1982.

12182 - **MUNNS, R., GREENWAY, H., KIRST, G.O.:** Halotolerant eucaryotes. - In: LANGE, O.L., NOBEL, P.S., OSMOND, C.B., ZIEGLER, H. (ed.): Physiological Plant Ecology III. Responses to the Chemical and Biological Environment. Pp. 59-135. Springer-Verlag, Berlin - Heidelberg - New York 1983.

12183 - **MUNNS, R., GREENWAY, H., SETTER, T.L., KUO, J.:** Turgor pressure, volumetric elastic modulus, osmotic volume and ultrastructure of *Chlorella emersonii* grown at high and low external NaCl. - J. exp. Bot. 34: 144-155, 1983.

*12184 - **MUNNS, R.E., PASSIOURA, J.B., RICHARDS, R.A.:** Plants and salinity. - In: CSIRO, Division of Plant Industry Report 1981-82. Pp. 19-22. CSIRO, Canberra 1982.

*12185 - **MURATA, Y., TANAKA, A.:** Agricultural aspects. - Recent Progress nat. Sci. Jap. 7: 79-87, 1982.

12186 - **MURRAY, F.:** Response of grapevines to fluoride under field conditions. - J. Amer. Soc. hort. Sci. 108: 526-529, 1983.

*12187 - **MUSICK, J.T., DUSEK, D.A.:** Irrigated corn yield response to water. - Trans. ASAE 23: 92-98, 103, 1980.

12188 - **MUTSAERS, H.J.W.:** Leaf growth in cotton (*Gossypium hirsutum* L.). 2. The influence of temperature, light, water stress and root restriction on the growth and initiation of leaves. - Ann. Bot. 51: 521-529, 1983.

12189 - **MYERS, S.C., FERREE, D.C.:** Influence of summer pruning and tree orientation on net photosynthesis, transpiration, shoot growth, and dry-weight distribution in young apple trees. - J. Amer. Soc. hort. Sci. 108: 4-9, 1983.

12190 - **NABORS, M.W.:** Increasing the salt and drought tolerance of crop plants. - In: RANDALL, D.D., BLEVINS, D.G., LARSON, R.L., RAPP, B.J., PALLARDY, S.G., NELSON, C.J., POLACCO, J.C., GEORGE, M.F. (ed.): Current Topics in Plant Biochemistry and Physiology, 1983. Volume 2. Pp. 165-184. University of Missouri, Columbia 1983.

12191 - **NAGARAJAH, S., SCHULZE, E.-D.:** Responses of *Vigna unguiculata* (L.) Walp. to atmospheric and soil drought. - Aust. J. Plant Physiol. 10: 385-394, 1983.

*12192 - **NAGY, A.H., HORÁNSZKY, A.:** Productivity and photosynthetic flexibility in some species of a grassland community. - Acta bot. Acad. Sci. Hung. 26: 389 - 395, 1980.

12193 - **NAIDOO, G.:** Effects of flooding on leaf water potential and stomatal resistance in *Bruguiera gymnorrhiza* (L.) Lam. - New Phytol. 93: 369-376, 1983.

12194 - **NAIK, G.R., JOSHI, G.V.:** Ineffectual role of proline metabolism in salt--stressed sugarcane leaves. - Proc. Ind. Acad. Sci., Plant Sci. 92: 265-269, 1983.

12195 - **NAMBIAR, P.T.C., NIGAM, S.N., DART, P.J., GIBBONS, R.W.:** Absence of root hairs in non-nodulating groundnut, *Arachis hypogaea* L. - J. exp. Bot. 34: 484-488, 1983.

12196 - **NARASIMHALU, P., CHOO, T.M.:** Effect of potassium carbonate on drying of red clover under humid field conditions. - Z. Acker- Pflanzenbau 152: 159-163, 1983.

12197 - **NASS, H.G.:** Relationship of plant water content and winterhardiness of winter wheat and fall rye grown in Atlantic Canada. - Can. j. Plant Sci. 63: 67-71, 1983.

12198 - **NASYROV, Yu.S., RASULOV, B.Kh., ALIEV, K.A., ASROROV, K.A.:** Fotosintez i fotodykhanie v ontogeneze khlopchatnika. [Photosynthesis and photorespiration during ontogeny of cotton leaf.] - Fiziol. Biokhim. kul't. Rast. 15: 43-46, 1983. [In R, ab: E.]

12199 - **NÁTR, L.:** Genotypic differences in growth and photosynthesis of young barley plants. - In: SARIĆ, M.R., LOUGHMAN, B.C. (ed.): Genetic Aspects of Plant Nutrition. Pp. 269-277. Martinus Nijhoff Publishers, The Hague - Boston - Lancaster 1983.

12200 - **NAUMENKO, A.I., KIRPA, N.Ya., ZOLOTOV, V.I., PONOMARENKO, A.K.:** Vlazhnost' i kachestvo semyan gibridov kukuruzy razlichnoĭ skorospelosti. [Moisture content and quality of seed in corn hybrids of various earliness.] - Sel'-skokhoz. Biol. 1983 (10): 40-43, 1983. [In R, ab: E.]

12201 - **NAVARA, J., KOZINKA, V.:** Die Ermittlung des Gesamtwasserverbrauches von Holzpflanzen mittels Kompensationslysimeter. - Biológia (Bratislava) 38: 39-50, 1983.

12202 - **NAYEK, A., BISWAS, A.K., Jr., CHOUDHURI, M.A.:** Effect of calcium on water--stress-induced biochemical changes and yield of field-grown rice. - Biol. Plant. 25: 117-123, 1983.

12203 - **NAZRUL-ISLAM, A.K.M.:** Tissue water relations in a range of plant species and its ecological significance. - Flora 174: 329-337, 1983.

12204 - **NEALES, T.F., FRASER, M.S., ROKSANDIC, Z.:** Carbon isotope composition of the halophyte *Disphyma clavellatum* (Haw.) Chinnock (*Aizoaceae*), as affected by salinity. - Aust. J. Plant Physiol. 10: 437-444, 1983.

12205 - **NECHIPORENKO, G.A., RYBALOVA, B.A.:** Vodoobmen plodov kabachka pri sozrevanii. [Water exchange in the vegetable marrow fruit during ripening.] - Fiziol. Rast. 30: 286-290, 1983. [In R, ab: E.]

*12206 - **NEGISI, K., HAYASHIDA, Y., YAGI, K.:** [Diurnal and seasonal fluctuations of stem wood temperatures in standing *Pinus densiflora*, *Cryptomeria japonica* and *Metasequoia glyptostroboides* trees with special reference to daytime depressions.] - Bull. Tokyo Univ. Forests 72: 1-15, 1982. [In Jap, ab: E.]

12207 - **NEILD, R.E., LOGAN, J., CARDENAS, A.:** Growing season and phenological response of sorghum as determined from simple climatic data. - Agr. Meteorol. 30: 35-48, 1983.

12208 - **NEJIDAT, A., ITAI, C., ROTH-BEJERANO, N.:** Stomatal response to ATP mediated by phytochrome. - Physiol. Plant. 57: 367-370, 1983.

12209 - **NELL, T.A., BARRETT, J.E., STAMPS, R.H.:** Water relations and frond curl of cut leatherleaf fern. - J. Amer. Soc. hort. Sci. 108: 516-519, 1983.

12210 - **NELSON, D.C., ANDERSON, R.C.:** Factors related to the distribution of prairie plants along a moisture gradient. - Amer. Midland Natur. 109: 367-375, 1983.

12211 - **NELSON, O.E.:** Genetics and plant breeding in relation to stress tolerance. - In: RAPER, C.D., Jr., KRAMER, P.J. (ed.): Crop Reactions to Water and Temperature Stresses in Humid, Temperate Climates. Pp. 351-357. Westview Press, Boulder 1983.

12212 - **NELSON, R.B., DAVIS, D.W., PALTA, J.P., LAING, D.R.:** Measurement of soil waterlogging tolerance in *Phaseolus vulgaris* L.: a comparison of screening techniques. - Scientia Hort. 20: 303-313, 1983.

12213 - **NEMEC, S.:** Oxygen, temperature, and water potential in shallow and deep soils of a citrus grove with blight. - Soil Crop Sci. Soc. Florida Proc. 42: 85-90, 1983.

12214 - **NEUMANN, L.:** Druckaufnehmer-Tensiometer mit kurzer Ansprechzeitkonstante zur automatischen Registrierung der Bodensaugspannung. - Arch. Acker- Pflanzenbau Bodenk. 27: 89-98, 1983.

12215 - **NEWMAN, E.I.:** Interactions between plants. - In: LANGE, O.L., NOBEL, P.S., OSMOND, C.B., ZIEGLER, H. (ed.): Physiological Plant Ecology III. Responses to the Chemical and Biological Environment. Pp. 679-710. Springer-Verlag, Berlin - Heidelberg - New York 1983.

12216 - **NEWMAN, K.W.:** Predicting depth to the water table based on species presence on an Oregon coast deflation plain. - Can. J. Bot. 61: 482-487, 1983.

12217 - **NEWTON, R.W., HEILMAN, J.L., VAN BAVEL, C.H.M.:** Integrating passive microwave measurements with a soil moisture/heat flow model. - Agr. Water Manage. 7: 379-389, 1983. Also in: STONE, J.F., WILLIS, W.O. (ed.): Plant Production and Management under Drought Conditions. Pp. 379-389. Elsevier Scientific Publishing Company, Amsterdam 1983.

*12218 - **NG, P.A.P., JARVIS, P.G.:** Hysteresis in the response of stomatal conductance in *Pinus sylvestris* L. needles to light: observations and a hypothesis. - Plant Cell Environ. 3: 207-216, 1980.

12219 - **NIEMANN, E.-J., RUHE, W.:** Die effektive Evapotranspiration als Parameter des Wasserhaushaltes in einem semiariden Repräsentativgebiet in den "Sierras de Córdoba" (Argentinien). - Z. Kulturtech. Flurbereinigung 24: 374-385, 1983.

12220 - **NIKOLAEV, E.V., SAPLEV, A.V., TUNIK, G.N.:** Znachenie rezhimov orosheniya i udobreniĭ dlya deyatel'nosti korneĭ, produktivnosti i kachestva zerna ozimoĭ pshenitsy. [Significance of irrigation and fertilization conditions for the activity of roots and quality of winter wheat grain.] - Fiziol. Biokhim. kul't. Rast. 15: 447-451, 1983. [In R, ab: E.]

*12221 - **NIKOLAEV, G.M., TAĬMLA, Ė.A., TUKEEVA, M.I., MOROZOVA, E.M., ZUBAREV, V.A., ERMAKOV, I.P.:** Issledovanie sostoyaniya vody v zrelykh pyl'tsevykh zernakh *Tradescantia paludosa.* [A study of water state in mature pollen grains of *Tradescantia paludosa.*] - Fiziol. Rast. 29: 312-316, 1982. [In R, ab: E.]

12222 - **NIKOLAEV, I.N., FROLOV, E.N., KONONENKO, A.A., RUBIN, A.B., GOL'DANSKIĬ, V. I.:** Issledovanie vnutrimolekulyarnoĭ podvizhnosti khromatoforov iz *Rhodopseudomonas sphaeroides* metodom messbauerovskoĭ spektroskopii. [Mössbauer spectroscopy of intramolecular mobility in chromatophores from *Rhodopseudomonas sphaeroides.*] - Biofizika 28: 131-133, 1983. [In R, ab: E.]

12223 - **NILIMA, MALIK, C.P.:** Physiological and biochemical changes under water stress in pigeonpea plants grown from seeds soaked in water and chemicals. - Ind. J. agr. Sci. 53: 550-556, 1983.

12224 - **NILSEN, E.T., SHARIFI, M.R., RUNDEL, P.W., JARRELL, W.M., VIRGINIA, R.A.:** Diurnal and seasonal water relations of the desert phreatophyte *Prosopis glandulosa* (honey mesquite) in the Sonoran Desert of California. - Ecology 64: 1381-1393, 1983.

12225 - **NOBEL, P.S.:** Low and high temperature influences on cacti. - In: MARCELLE, R., CLIJSTERS, H., VAN POUCKE, M. (ed.): Effects of Stress on Photosynthesis. Pp. 165-174. Martinus Nijhoff / Dr W. Junk Publishers, The Hague - Boston - London 1983.

12226 - **NOBEL, P.S.:** Nutrient levels in cacti - relation to nocturnal acid accumulation and growth. - Amer. J. Bot. 70: 1244-1253, 1983.

12227 - **NOBEL, P.S., HARTSOCK, T.L.:** Relationships between photosynthetically active radiation, nocturnal acid accumulation, and CO_2 uptake for Crassulacean acid metabolism plant, *Opuntia ficus-indica.* - Plant Physiol. 71: 71-75, 1983.

12228 - **NOBEL, P.S., JORDAN, P.W.:** Transpiration stream of desert species: resistances and capacitances for a C_3, a C_4 and a CAM plant. - J. exp. Bot. 34: 1379-1391, 1983.

12229 - **NORBY, R.J., LUXMOORE, R.J.:** Growth analysis of soybean exposed to simulated acid rain and gaseous air pollutants. - New Phytol. 95: 277-287, 1983.

12230 - **NUKAYA, A., MASUI, M., ISHIDA, A.:** Salt tolerance of muskmelons as affected by various salinities in nutrient solution culture. - J. Jap. Soc. hort. Sci. 52: 167-173, 1983.

12231 - **NUKAYA, A., MASUI, M., ISHIDA, A.:** Salt tolerance of muskmelons at different growth stages as affected by diluted sea water. - J. Jap. Soc. hort. Sci. 52: 286-293, 1983.

12232 - **NUKAYA, A., MASUI, M., ISHIDA, A.:** Salt tolerance of muskmelons as affected by various salinities in sand culture. - J. Jap. Soc. hort. Sci. 51: 427-434, 1983.

12233 - **NULTSCH, W., GRAHLE, A.:** Mikroskopisch-Botanisches Praktikum für Anfänger. 7., Überarbeitete Auflage. - Georg Thieme Verlag, Stuttgart 1983.

*12234 - **NUNES, M.A., CORREIA, O.C.A.:** Water relations and gas exchanges in *Ceratonia siliqua* L. - Port. Acta biol., Ser. A, 16: 151-163, 1980.

12235 - **NYMAN, L.P., GONZALES, C.J., ARDITTI, J.:** *In-vitro* selection for salt tolerance of taro (*Colocasia esculenta* var. *antiquorum*). - Ann. Bot. 51: 229-236, 1983.

12236 - **O'BRIEN, P.J., ALLEN, E.J., BEAN, J.N., GRIFFITH, R.L., JONES, S.A., JONES, J.L.:** Accumulated day-degrees as a measure of physiological age and the relationships with growth and yield in early potato varieties. - J. agr. Sci. 101: 613-631, 1983.

12237 - **OFFLER, C.E., PATRICK, J.W., ROSE, R.J.:** Chloroplast characteristics and photochemical efficiency of low irradiance grown *Eucalyptus pilularis* Sm. seedlings. - Photosynthetica 17: 227-234, 1983.

12238 - **O'HARA, E.P., TOM, R.D., MOORE, T.A.:** Determination of the *in vivo* absorption and photosynthetic properties of the lichen *Acarospora schleicheri* using photoacoustic spectroscopy. - Photochem. Photobiol. 38: 709-715, 1983.

12239 - **OJALA, J.C., JARRELL, W.M., MENGE, J.A., JOHNSON, E.L.V.:** Influence of mycorrhizal fungi on the mineral nutrition and yield of onion in saline soil. - Agron. J. 75: 255-259, 1983.

12240 - **OKADA, M., AIBA, S.:** Simulation of water-bloom in a eutrophic lake - II. Reassessment of buoyancy, gas vacuole and turgor pressure of *Microcystis aeruginosa*. - Water Res. 17: 877-882, 1983.

12241 - **OKADA, M., AIBA, S.:** Simulation of water-bloom in a eutrophic lake - III. Modeling the vertical migration and growth of *Microcystis aeruginosa*. - Water Res. 17: 883-893, 1983.

12242 - **OLABIYI, O., KAISER, W.M., HEBER, U.:** Respiratory activities of plant mitochondria under osmotic stress. - Z. Pflanzenphysiol. 111: 155-167, 1983.

12243 - **OLADELE, F.A.:** Scanning electron microscope study of stomatal-complex configuration in *Cupressaceae*. - Can. J. Bot. 61: 1232-1240, 1983.

*12244 - **OLOVYANNIKOVA, I.N.:** Potreblenie vlagi vyazom melkolistnym v polupustyne Prikaspiya pri razlichoĭ vlagoobespechennosti. [Moisture consumption by Chinese elm with different moisture supply in the Caspian semidesert.] - Pochvovedenie 1980(4): 92-102, 1980. [In R, ab: E.]

12245 - **OLSEN, M.W., MISAGHI, I.J., GOLDSTEIN, D., HINE, R.B.:** Water relations in cotton plants infected with *Phymatotrichum*. - Phytopathology 73: 213-216, 1983.

12246 - **OLSSON, K.A., MILTHORPE, F.L.:** Diurnal and spatial variation in leaf water potential and leaf conductance of irrigated peach trees. - Aust. J. Plant Physiol. 10: 291-298, 1983.

*12247 - **OMASA, K., ABO, F., AIGA, I., HASHIMOTO, Y.:** Image instrumentation of plants exposed to air pollutants - quantification of physiological information included in thermal infrared images. - Trans. Soc. Instrument Cont. Eng. 17: 657-664, 1981.

*12248 - **OMASA, K., AIGA, I., HASHIMOTO, Y.:** Image instrumentation for evaluating the effects of air pollutants on plants. - In: IMEKO 9th World Congress. Volume 3. Pp. 308-317. Akadémiai Kiado, Budapest 1982.

*12249 - **OMASA, K., HASHIMOTO, Y., AIGA, I.:** A quantitative analysis of the relationships between SO_2 or NO_2 sorption and their acute effects on plant leaves using image instrumentation. - Environ. Cont. Biol. 19: 59-67, 1981.

*12250 - **OMASA, K., HASHIMOTO, Y., AIGA, I.**: A quantitative analysis of the relationships between O_3 sorption and its acute effects on plant leaves using image instrumentation. - Environ. Cont. Biol. 19: 85-92, 1981.

12251 - **OMASA, K., HASHIMOTO, Y., AIGA, I.**: Observation of stomatal movements of intact plants using an image instrumentation system with a light microscope. - Plant Cell Physiol. 24: 281-288, 1983.

12252 - **OMER, L.S., HORVATH, S.M., SETARO, F.**: Salt regulation and leaf senescence in aging leaves of *Jaumea carnosa* (Less.) Gray (*Asteraceae*), a salt Marsh species exposed to NaCl stress. - Amer. J. Bot. 70: 363-368, 1983.

*12253 - **O'NEAL, S.W., PRINCE, J.S.**: Relationship between seasonal growth, photosynthetic production and apex mortality of *Caulerpa paspaloides* (*Chlorophyceae*). - Mar. Biol. 72: 61-67, 1982.

12254 - **O'NEIL, K.J., CARROW, R.N.**: Perennial ryegrass growth, water use, and soil aeration status under soil compaction. - Agron. J. 75: 177-180, 1983.

12255 - **O'NEILL, M.K., HOFMANN, W., DOBRENZ, A.K., MARCARIAN, V.**: Drought response of sorghum hybrids under a sprinkler irrigation gradient system. - Agron. J. 75: 102-107, 1983.

*12256 - **O'NEILL, S.D., PRIESTLEY, D.A., CHABOT, B.F.**: Temperature and aging effects on leaf membranes of a cold hardy perennial, *Fragaria virginiana*. - Plant Physiol. 68: 1409-1415, 1981.

12257 - **ONG, C.K.**: Response to temperature in a stand of pearl millet (*Pennisetum typhoides* S. & H.). 1. Vegetative development. - J. exp. Bot. 34: 322-336, 1983.

12258 - **OOSTERHUIS, D.M., CARTWRIGHT, P.M.**: Spike differentiation and floret survival in semidwarf spring wheat as affected by water stress and photoperiod. - Crop Sci. 23: 711-717, 1983.

12259 - **OOSTERVELD, M., NICHOLAICHUK, W.**: Water requirements, availability and development restraints for increased crop productivity in Canada. - Can. J. Plant Sci. 63: 33-44, 1983.

12260 - **ÖQUIST, G.**: Effects of low temperature on photosynthesis. - Plant Cell Environ. 6: 281-300, 1983.

12261 - **ORT, D.R., MARTIN, B.**: Chilling-induced inhibition of photosynthesis in tomato. - In: MARCELLE, R., CLIJSTERS, H., VAN POUCKE, M. (ed.): Effects of Stress on Photosynthesis. Pp. 227-235. Martinus Nijhoff / Dr W. Junk Publishers, The Hague - Boston - London 1983.

12262 - **ORTH, C.**: Einfluss der Wasserversorgung auf die Assimilattranslokation bei Reben. - Thesis, Institut für Obst-, Gemüse- und Weinbau, Universität Hohenheim, Stuttgart-Hohenheim 1983.

12263 - **OSHMARINA, V.I., SHAMINA, Z.B., SHEVYAKOVA, N.I.**: Kharakteristika rezistentnykh k NaCl i ėtioninu kletochnykh liniĭ *Nicotiana sylvestris*. [Characteristics of *Nicotiana sylvestris* cell lines resistant to NaCl and ethionine.] - Fiziol. Rast. 30: 527-536, 1983. [In R, ab: E.]

12264 - **OSMOND, C.B.**: Interactions between irradiance, nitrogen nutrition, and water stress in the sun-shade responses of *Solanum dulcamara*. - Oecologia 57: 316-321, 1983.

*12265 - **OSOTSAPAR, Y., MERCADO, B.T.**: Osmotic adjustment in salt-stressed rice plants. - Philippine J. Sci. 107: 51-60, 1978.

12266 - **O'TOOLE, D.K.:** Weighing technique for determining bacterial dry mass based on rate of moisture uptake. - Appl. environ. Microbiol. 46: 506-508, 1983.

12267 - **O'TOOLE, J.C., HATFIELD, J.L.:** Effect of wind on the crop water stress index derived by infrared thermometry. - Agron. J. 75: 811-817, 1983.

12268 - **OUTLAW, W.H., Jr.:** Current concepts on the role of potassium in stomatal movements. - Physiol. Plant. 59: 302-311, 1983.

*12269 - **OUTLAW, W.H., Jr., MANCHESTER, J., BROWN, P.H.:** High levels of malic enzyme activities in *Vicia faba* L. epidermal tissue. - Plant Physiol. 68: 1047-1051, 1981.

*12270 - **OVERDIECK, D., FAENSEN-THIEBES, A.:** Wirkungen von Ozon auf CO_2-Gaswechsel und Transpiration bei *Nicotiana tabacum* L. Bel W-3 und *Phaseolus vulgaris* L. var. *saxa*. - Verhandl. Ges. Ökol. 9: 269-275, 1981.

12271 - **OWEN, I.G., WILMAN, D.:** Differences between grass species and varieties in rate of drying at 25 °C. - J. agr. Sci. 100: 629-636, 1983.

12272 - **OWNBEY, R.S., MAHALL, B.E.:** Salinity and root conductivity: differential responses of a coastal succulent halophyte, *Salicornia virginica*, and a weedy glycophyte, *Raphanus sativus*. - Physiol. Plant. 57: 189-195, 1983.

12273 - **PAEZ, A., HELLMERS, H., STRAIN, B.R.:** CO_2 enrichment, drought stress and growth of Alaska pea plants (*Pisum sativum*). - Physiol. Plant. 58: 161-165, 1983.

12274 - **PAHLICH, E., KERRES, R., JÄGER, H.-J.:** Influence of water stress on the vacuole/extravacuole distribution of proline in protoplasts of *Nicotiana rustica*. - Plant Physiol. 72: 590-591, 1983.

12275 - **PALATINUS, Ľ.:** Šírenie hrdzí (*P. graminis, P. recondita* a *P. striiformis*) na ozimnej pšenici pri závlahe. [The spreading of rusts (*P. graminis, P. recondita* a *P. striiformis*) on winter wheat in irrigated fields.] - Sborník ÚVTIZ - Ochrana Rost. (Praha) 19: 17-23, 1983. [In Slov, ab: R, E.]

12276 - **PALIN, M.A., PETTY, J.A.:** Permeability to water of the wood cell wall and its variation with temperature. - Wood Sci. Technol. 17: 187-193, 1983.

12277 - **PALLARDY, S.G., PARKER, W.C., WHITEHOUSE, D.L., HINCKLEY, T.M., TESKEY, R. O.:** Physiological responses to drought and drought adaptation in woody species. - In: RANDALL, D.D., BLEVINS, D.G., LARSON, R.L., RAPP, B.J., PALLARDY, S.G., NELSON, C.J., POLACCO, J.C., GEORGE, M.F. (ed.): Current Topics in Plant Biochemistry and Physiology, 1983. Volume 2. Pp. 185-199. University of Missouri, Columbia 1983.

12278 - **PALTA, J.A.:** Evaluation of the hydraulic press for measurement of leaf water potentials in cassava. - J. agr. Sci. 101: 407-410, 1983.

12279 - **PALTA, J.A.:** Photosynthesis, transpiration, and leaf diffusive conductance of the cassava leaf in response to water stress. - Can. J. Bot. 61: 373-376, 1983.

12280 - **PALTA, J.P., LEE-STADELMANN, O.Y.:** Vacuolated plant cells as ideal osmometer: reversibility and limits of plasmolysis, and estimation of protoplasm volume in control and water-stress-tolerant cells. - Plant Cell Environ. 6: 601-610, 1983.

12281 - **PAMMENTER, N.W., SMITH, V.R.:** The effect of salinity on leaf water relations and chemical composition in the sub-Antarctic tussock grass *Poa cookii* Hook F. - New Phytol 94: 585-594, 1983.

12282 - **PAPADOPOULOS, I., RENDIG, V.V.:** Interactive effects of salinity and nitrogen on growth and yield of tomato plants. - Plant Soil 73: 47-57, 1983.

12283 - **PAPADOPOULOS, I., RENDIG, V.V.:** Tomato plant response to soil salinity. - Agron. J. 75: 696-700, 1983.

12284 - **PAPP, J.C., BALL, M.C., TERRY, N.:** A comparative study of the effects of NaCl salinity on respiration, photosynthesis, and leaf extension growth in *Beta vulgaris* L. (sugar beet). - Plant Cell Environ. 6: 675-677, 1983.

*12285 - **PARAMESWARA, G., KRISHNASASTRY, K.S.:** Variability in leaf elongation rate and reduction in green leaf length in sorghum genotypes under moisture stress and on alleviation of stress. - Ind. J. agr. Sci. 52: 102-106, 1982.

12286 - **PARKE, J.L., LINDERMAN, R.G., BLACK, C.H.:** The role of ectomycorrhizas in drought tolerance of Douglas-fir seedlings. - New Phytol. 95: 83-95, 1983.

12287 - **PARKINSON, K.J., DAY, W.:** The influence of water stress on photosynthesis in a barley crop. - In: MARCELLE, R., CLIJSTERS, H., VAN POUCKE, M. (ed.): Effects of Stress on Photosynthesis. Pp. 65-74. Martinus Nijhoff / Dr W. Junk Publishers, The Hague - Boston - London 1983.

*12288 - **PARLANGE, J.-Y., BRADDOCK, R.D., SIMPSON, R.W., HILL, D.E.:** Optimization principle for air and water movement. - Soil Sci. 133: 4-9, 1982.

12289 - **PARRISH, D.J., WOLF, D.D.:** Kinetics of tall fescue leaf elongation: Responses to changes in illumination and vapor pressure. - Crop Sci. 23: 659-663, 1983.

*12290 - **PARSONS, L.R., WHEATON, T.A., WHITNEY, J.D.:** Low volume microsprinkler under-tree irrigation for frost protection of young citrus trees. - Proc. Florida State hort. Soc. 94: 55-59, 1981.

12291 - **PASSIOURA, J.B.:** Roots and drought resistance. - Agr. Water Manage. 7: 265-280, 1983. Also in: STONE, J.F., WILLIS, W.O. (ed.): Plant Production and Management under Drought Conditions. Pp. 265-280. Elsevier Scientific Publishing Company, Amsterdam 1983.

*12292 - **PASUTHOVÁ, J.:** Vodní provoz a rezistence smrku. [Water management of norway spruce and its resistance.] - Práce VÚLHM (Praha) 60: 99-109, 1982. [In Czech, ab: R, E.]

12293 - **PÁSZTOR, K., PEPÓ, P., PEPÓ, P., EGRI, K.:** Mutáns vonalakkal előállított hibridkombinációk vízleadó képességének értékelése matematikai-statisztikai módszerekkel. [Water release by hybrid combinations of maize produced with mutant lines as evaluated by mathematical-statistical methods.] - Növénytermelés 32: 1-12, 1983. [In Hung, ab: E.]

12294 - **PATAKY, J.K., BEUTE, M.K.:** Effects of inoculum burial, temperature, and soil moisture on survival of *Cylindrocladium crotalariae* microsclerotia in North Carolina. - Plant Dis. 67: 1379-1382, 1983.

12295 - **PATEL, C.L., PADALIA, M.R., BABARIA, N.B.:** Growth and plant-water relation in groundnut grown under different soil-moisture stresses. - Ind. J. agr. Sci. 53: 340-345, 1983.

12296 - **PATEL, C.S., MANDAL, A.K.:** Effect of moisture regimes and level of fertilizer application on yield and water requirement of jute (*Corchorus olitorius* L. and *Corchorus capsularis* L.). - J. agr. Sci. 101: 311-316, 1983.

*12297 - **PATEL, J.C., SINGH, R.M.:** Effect of irrigation, mulching and bioregulator on the productivity of sunflower. - Ind. J. Agron. 25: 122-128, 1980.

*12298 - **PATEL, J.J., PATEL, M.S., KANZARIA, M.V.:** Effect of lime and sulphur on iron availability under different soil moisture regimes. - Ind. J. agr. Res. 16: 83-86, 1982.

12299 - **PATERSON, D.R., PATERSON, E.W., KENWORTHY, J.B.:** Physiological studies on temperate lichen species. I. A mathematical model to predict assimilation in the field, based on laboratory responses. - New Phytol. 94: 605-618, 1983.

12300 - **PATTERSON, D.T.:** Phenotypic and physiological comparisons of field and phytotron grown plants. - In: RAPER, C.D., Jr., KRAMER, P.J. (ed.): Crop Reactions to Water and Temperature Stresses in Humid, Temperate Climates. Pp. 299-315. Westview Press, Boulder 1983.

12301 - **PATTERSON, D.T., FLINT, E.P.:** Comparative water relations, photosynthesis, and growth of soybean (*Glycine max*) and seven associated weeds. - Weed Sci. 31: 318-323, 1983.

*12302 - **PÄTZOLD, H., BAUER, U.:** Účinnost závlahy postřikem čistou vodou na výnos, kontinuitu produkce a jakost píce různých travních porostů v severních oblastech NDR. [Efficiency of sprinkler irrigation in yield, continuity of production and quality of fodder of various grass stands in northern G.D.R.] - Acta Univ. Agr., Fac. agron. (Brno) A28(3-4): 275-283, 1980. [In Czech, ab: G.]

12303 - **PAVLIK, B.M.:** Nutrient and productivity relations of the dune grasses *Ammophila arenaria* and *Elymus mollis* I. Blade photosynthesis and nitrogen use efficiency in the laboratory and field. - Oecologia 57: 227-232, 1983.

12304 - **PAW U, K.T., GUEYE, M.:** Theoretical and measured evaporation rates from an exposed Piche atmograph. - Agr. Meteorol. 30: 1-11, 1983.

12305 - **PAZOUREK, J., NÁTR, L.:** Kvantitativní anatomická heterogenita listů ječmene a genotypové rozdíly. [Quantitative anatomical heterogeneity of barley leaves and genotypic differences.] - In: Dny Rostlinné Fyziologie III. Pp. 257-259. ČSBS, Praha 1983. [In Czech.]

12306 - **PEARCY, R.W.:** Physiological consequences of cellular water deficits: nonstomatal inhibition of photosynthesis by water stress. - In: TAYLOR, H.M., JORDAN, W.R., SINCLAIR, T.R. (ed.): Limitations of Efficient Water Use in Crop Production. Pp. 277-287. Amer. Soc. Agron., Inc., Crop Sci. Soc. Amer., Inc., Soil Sci. Soc. Amer., Inc., Madison 1983.

12307 - **PEARCY, R.W., CALKIN, H.W.:** Carbon dioxide exchange of C_3 and C_4 tree species in the understory of a Hawaiian forest. - Oecologia 58: 26-32, 1983.

12308 - **PEAT, W.E.:** Developmental physiology. - In: HEBBLETHWAITE, P.D. (ed.): The Faba Bean (*Vicia faba* L.). A Basis for Improvement. Pp. 103-132. Butterworths, London - Boston - Durban - Singapore - Sydney - Toronto - Wellington 1983.

12309 - **PEDRO, M.J., Jr., VILLA NOVA, N.A.:** Estimativa de resistência ao transporte de vapor d'água em cultura de soja. [Estimation of the resistance to water vapour transport in soybean crop.] - Pesq. agropec. Bras. 18: 691-694, 1983. [In Port, ab: E.]

*12310 - **PEMADASA, M.A.:** Abaxial and adaxial stomatal behaviour and responses to fusicoccin on isolated epidermis of *Commelina communis* L. - New Phytol. 89: 373-384, 1981.

12311 - **PEMADASA, M.A.:** Effects of benzo-18-crown-6 on abaxial and adaxial stomatal opening and its antagonism with abscisic acid. - New Phytol. 93: 13-24, 1983.

12312 - **PEMADASA, M.A.:** Sodium ions can eliminate the normal disparity in abaxial and adaxial stomatal opening on isolated epidermes. - New Phytol. 94: 201-209, 1983.

12313 - **PEMADASA, M.A.:** Effects of KCl concentration on changes in starch and malate levels in abaxial and adaxial epidermis in relation to stomatal opening. - J. exp. Bot. 34: 514-522, 1983.

12314 - **PENKA, M.:** Vodní provoz heterotrofních rostlin. [Water regime of heterotrophic plants.] - Lesnictví (Praha) 29: 961-978, 1983. [In Czech, ab: R, E.]

12315 - **PENKA, M., ČERMÁK, J., KUČERA, J., PRAX, A., ÚLEHLA, J., ŽÍDEK, V.:** Spotřeba vody u dubu (*Quercus robur* L.) v aluviu řeky Dyje za nelimitujících vlhkostních podmínek. [Water consumption in oak (*Quercus robur* L.) in the alluvium of the Dyje river in nonlimiting moisture conditions.] - Lesnictví (Praha) 29: 481-196, 1983. [In Czech, ab: R, E.]

12316 - **PERRY, S.W., KRIEG, D.R., HUTMACHER, R.B.:** Photosynthetic rate control in cotton. Photorespiration. - Plant Physiol. 73: 662-665, 1983.

12317 - **PESCHL, A.:** CO_2-Gaswechsel und Wasserhaushalt junger Pflanzen von *Pseudotsuga menziesii* und *Abies grandis* bei Frosttrocknis. - Flora 173: 33-47, 1983.

12318 - **PETERSON, C.A., EMANUEL, M.E.:** Casparian strips occur in onion root hypodermal cells: Evidence from band plasmolysis. - Ann. Bot. 51: 135-137, 1983.

12319 - **PETTY, J.A., PALIN, M.A.:** Permeability to water of the fibre cell wall material of two hardwoods. - J. exp. Bot. 34: 688-693, 1983.

12320 - **PEW, W.D., GARDNER, B.R.:** Effects of irrigation practices on vine growth, yield, and quality of muskmelons. - J. Amer. Soc. hort. Sci. 108: 134-137, 1983.

*12321 - **PEZESHKI, S.R., HINCKLEY, T.M.:** The stomatal response of red alder and black cotton-wood to changing water status. - Can. J. Forest Res. 12: 761-771, 1982.

*12322 - **PFEIFFER, T., SCHRADER, L.E., BINGHAM, E.T.:** Physiological comparisons of isogenic diploid-tetraploid, tetraploid-octoploid alfalfa populations. - Crop Sci. 20: 299-303, 1980.

12323 - **PHAM THI, A.T., BORREL-FLOOD, C., JUSTIN, A.-M., MAZLIAK, P., VIEIRA da SILVA, J.:** Action de la sécheresse sur l'incorporation du 1-^{14}C-acétate dans les lipides de feuilles de Cotonnier (*Gossypium hirsutum* L.). - C. R. Acad. Sci. Paris, Sér. III 297: 613-616, 1983.

12324 - **PHAM THI, A.T., FERRARI-ILIOU, R., VIEIRA da SILVA, J.:** Effects of water stress on the fatty acid and lipid composition of cotton chloroplasts. - In: Sixth International Congress on Photosynthesis. Abstracts. Volume 1. Pp. 399. Brussels 1983.

*12325 - **PHILLIPS, S.A., HUTCHINSON, R.L., BARTLESON, J.L., TALBOT, T.:** Irrigation boosts cotton, soybean yields in Macon Ridge tests. - Louisiana Agr. 24 (3): 22-23, 1981.

12326 - **PINTER, P.J., Jr., FRY, K.E., GUINN, G., MAUNEY, J.R.:** Infrared thermometry: A remote sensing technique for predicting yield in water-stressed cotton. - Agr. Water Manage. 6: 385-395, 1983.

*12327 - **PINTER, P.J., Jr., REGINATO, R.J.:** A thermal infrared technique for monitoring cotton water stress and scheduling irrigations. - Trans. ASAE 25: 1651-1655, 1982.

12328 - **PITMAN, W.D., HOLT, E.C., CONRAD, B.E., BASHAW, E.C.:** Histological differences in moisture-stressed and nonstressed kleingrass forage. - Crop Sci. 23: 793-795, 1983.

12329 - **PLESKANKA, J., POSPÍŠILOVÁ, J., SOLÁROVÁ, J., TICHÁ, I., ČATSKÝ, J.:** Response of epidermal conductance to irradiance as affected by changes in carbon dioxide concentration, leaf water potential or age. Mathematical approximation. - Photosynthetica 17: 161-167, 1983.

12330 - **POCHET, P.:** Quelques paramètres foliaires en relation avec la productivité chez le théier. - Café Cacao Thé 27: 177-182, 1983.

12331 - **PODOLÁK, M.:** Vplyv hnojív a závlahy na úrodu kukurice na siláž. [The effect of fertilizers and irrigation on silage maize yield.] - Rost. Výroba (Praha) 29: 857-864, 1983. [In Slov, ab: R, E.]

12332 - **POLONENKO, D.R., DUMBROFF, E.B., MAYFIELD, C.I.:** Microbial responses to salt-induced osmotic stress. III. Effects of stress on metabolites in the roots, shoots and rhizosphere of barley. - Plant Soil 73: 211-225, 1983.

12333 - **PONDER, F., Jr.:** Soil moisture levels and mycorrhizal infection of black walnut seedlings. - Commun. Soil Sci. Plant Anal. 14: 507-511, 1983.

12334 - **PONDER, H.G., GILLIAM, C.H., WILKINSON, E.H.:** Response of container-grown photinia to moisture stress. - Scientia Hort. 21: 369-374, 1983.

*12335 - **POOLE, R.T., CONOVER, C.A.:** Influence of maximum air temperatures and irrigation frequencies during high temperature periods on growth of four foliage plants. - HortScience 16: 556-557, 1981.

*12336 - **POPOV, Ė.G., BEZDENEZHNYKH, V.A., TALANOV, A.V.:** Optimizatsiya protsessa formirovaniya produktivnosti teplichnoĭ kul'tury ogurtsa. [Optimization of the process of formation of productivity of a glasshouse cucumber crop.] - In: Vliyanie Faktorov Vneshneĭ Sredy i Fiziologicheski Aktivnykh Veshchestv na Termorezistentnost' i Produktivnost' Rasteniĭ. Pp. 116-126, 158. Inst. Biol. Karel'skogo Filiala Akad. Nauk SSSR, Petrozavodsk 1982. [In R.]

12337 - **POPOV, P.S., KOZHEVNIKOVA, V.N.:** Dinamika kachestvennogo sostava masla v semenakh podsolnechnika v svyazi s orosheniem. [Dynamics of oil quantitative composition in sunflower seeds in connection with irrigation.] - Fiziol. Biokhim. kul't. Rast. 15: 158-162, 1983. [In R, ab: E.]

*12338 - **POPOVA, L.P., DIMITROVA, O.D., VAKLINOVA, S.G.:** Effect of ABA on the chlorophyll content, the intensity of the photosynthetic CO_2 fixation and the activity of the carboxylating enzymes in C_3 and C_4 plants. - Dokl. Bolg. Akad. Nauk 35: 1291-1294, 1982.

12339 - **POROZOVA, O.A.:** Radiochuvstvitel'nost' semyan sosny v zavisimosti ot ikh vlazhnosti i srokov khraneniya posle oblucheniya. [Pine seeds radiosensitivity as depended upon their humidity and the term of storage after radiation exposure.] - Ėkologiya 3: 82-84, 1983. [In R.]

*12340 - **POSGAY, E.:** A vízellátás és a termés közötti kapcsolat az öntözéses növénytermesztében. IV. Etkezési szárazbab. [Relationship between water supply and yield in crop growing under irrigation. IV. Dried beans.] - Növénytermelés 31: 545-550, 1982. [In Hung, ab: E.]

12341 - **POSGAY, E.:** A vízellátás és a termés közötti kapcsolat az öntözéses növénytermesztében. V. A vízhiány és az öntözés jelzése szimulátorral. [Relationship between water supply and yield in crop growing under irrigation. V. Indication of water shortage and irrigation by a simulator.] - Növénytermelés 32: 339-346, 1983. [In Hung, ab: E.]

*12342 - **POSPÍŠILOVÁ, J., SOLÁROVÁ, J. (ed.)**: Water-in-Plants Bibliography. Volume 6 1980. References no. 6606 - 8128 / AAS - ZOB. - Dr W. Junk Publishers, The Hague - Boston - London 1982.

*12343 - **POSPÍŠILOVÁ, J., SOLÁROVÁ, J. (ed.)**: Water-in-Plants Bibliography. Volume 7 1981. References no. 8129 - 9676 / ABD - ZYA. - Dr W. Junk bv Publishers, The Hague - Boston - London 1982.

12344 - **POSPÍŠILOVÁ, J., SOLÁROVÁ, J. (ed.)**: Water-in-Plants Bibliography. Volume 8 1982. References no. 9677 - 11192 / ABD - ZYA. - Dr W. Junk Publishers, The Hague - Boston - London 1983.

12345 - **POSPÍŠILOVÁ, J., SOLÁROVÁ, J.**: Otevírání průduchů indukované zvýšením ozářenosti v závislosti na vodním potenciálu a stáří listu. [Light induced stomatal opening in dependence on water potential and age of leaf.] - In: Dny Rostlinné Fyziologie III. Pp. 262-263. ČSBS, Praha 1983. [In Czech.]

12346 - **POTVIN, M.A., WERNER, P.A.**: Water use physiologies of co-occurring goldenrods (*Solidago juncea* and *S. canadensis*): implications for natural distributions. - Oecologia 56: 148-152, 1983.

*12347 - **POWER, J.F.**: Response of semiarid grassland sites to nitrogen fertilization: I. Plant growth and water use. - Soil Sci. Soc. Amer. J. 44: 545-550, 1980.

12348 - **POWER, J.F.**: Soil management for efficient water use: soil fertility. - In: TAYLOR, H.M., JORDAN, W.R., SINCLAIR, T.R. (ed.): Limitations to Efficient Water Use in Crop Production. Pp. 461-470. Amer. Soc. Agron., Inc., Crop Sci. Soc. Amer., Inc., Soil Sci. Soc. Amer. Inc., Madison 1983.

12349 - **POWLES, S.B., BERRY, J.A., BJÖRKMAN, O.**: Interaction between light and chilling temperature on the inhibition of photosynthesis in chilling-sensitive plants. - Plant Cell Environ. 6: 117-123, 1983.

12350 - **PRANGE, R.K., ORMROD, D.P.**: Differential response in the water status of immature and mature fronds of the ostrich fern (*Matteuccia struthiopteris* L.) Todaro) to a mild water stress. - Plant Physiol. 72: 96-98, 1983.

12351 - **PRANGE, R.K., ORMROD, D.P., PROCTOR, J.T.A.**: Effect of water stress on gas exchange in fronds of the ostrich fern (*Matteuccia struthiopteris* (L.) Todaro). - J. exp. Bot. 34: 1108-1116, 1983.

*12352 - **PRICE, J.F., HARBAUGH, B.K., STANLEY, C.D.**: Response of mites and leafminers to trickle irrigation rates in spray *Chrysanthemum* production. - HortScience 17: 895-896, 1982.

*12353 - **PRICE, N.W.**: A comparison of water balance components in natural and plantation forests in El Salvador, Central America. - Turrialba 32: 399-416, 1982.

12354 - **PRIEHRADNÝ, S.**: Zmeny vodného sýtostného deficitu po infekcii múčnatkou rezistentného kultivaru jačmeňa. [Changes in the water saturation deficit following infection by powdery mildew of a resistant barley cultivar.] - Biologia (Bratislava) 38: 33-38, 1983. [In Slov, ab: E, R.]

*12355 - **PRIOUL, J.-L.**: Limiting factors in photosynthesis - from the chloroplast to the plant canopy. - In: HÉLÈNE, C., CHARLIER, M., MONTENAY-GARESTIER, T., LAUSTRIAT, G. (ed.): Trends in Photobiology. Pp. 633-643. Plenum Publishing Corp., New York - London 1982.

12356 - **PRIOUL, J.L., CORNIC, G.**: Discussion of stomatal and non stomatal components in photosynthesis decline under stress conditions. - In: Sixth International Congress on Photosynthesis. Abstracts. Volume 1. Pp. 400. Brussels 1983.

*12357 - **PROCTOR, M.C.F.**: Physiological ecology: Water relations, light and temperature responses, carbon balance. - In: SMITH, A.J.E. (ed.): Bryophyte Ecology. Pp. 333-381. Chapman & Hall, London - New York 1982.

12358 - **PROKOF'EV, A.A., KARIMOV, Kh.Kh., ASROROV, K.A., CHERNER, R.I., RASULOV, S., GADOIEV, N.**: Kruglogodichnoe ispol'zovanie oroshaemykh zemel' v khlopkoseyushchikh raĭonakh strany. [Year-round intensive use of irrigated lands in cotton-growing regions.] - Fiziol. Rast. 30: 813-820, 1983. [In R, ab: E.]

12359 - **PROKOF'EV, A.A., OBRUCHEVA, N.V., KOVADLO, L.S., KULIEVA, L.K., KOZHEMYAKINA, I.S.**: Kriticheskiĭ uroven' ovodnennosti semyan dlya nachala ikh prorastaniya. [The critical level of seed water content for initiation of germination.] - Fiziol. Rast. 30: 178-183, 1983. [In R, ab: E.]

12360 - **PROKOF'EV, A.A., RYBALOVA, B.A., NECHIPORENKO, G.A.**: Vodoobmen sozrevayushchikh sukhikh i sochnykh plodov maka maslichnogo i kabachka. [Water relations in ripening dry and fleshy fruits of oil poppy and vegetable marrow.] - Fiziol. Rast. 30: 1135-1142, 1983. [In R, ab: E.]

12361 - **PRUNTY, L.**: Soil water and population influence on hybrid sunflower yield and uniformity of stand. - Agron. J. 75: 745-749, 1983.

12362 - **PURVIS, A.C.**: Effects of film thickness and storage temperature on water loss and internal quality of seal-packaged grapefruit. - J. Amer. Soc. hort. Sci. 108: 562-566, 1983.

12363 - **PUSTOVOĬTOVA, T.N., EREMIN, G.V., RASSVETAEVA, Ė.G., ZHDANOVA, N.E.**: Osobennosti zasukhoustoĭchivosti plodovykh poliploidnykh rasteniĭ. [Drought resistance of polyploid plants.] - Sel'skokhoz. Biol. 1983 (3): 53-57, 1983. [In R, ab: E.]

*12364 - **QUAGLIETTA CHIARANDÀ, F., ZERBI, G.**: Effect of irrigation regimen on yield and water consumption of greenhouse tomato grown in lysimeters. - Acta Hort. 119: 179-190, 1981.

12365 - **QUAMME, H.A.**: Relationship of air temperature to water content and supercooling of overwintering peach flower buds. - J. Amer. Soc. hort. Sci. 108: 697-701, 1983.

*12366 - **QUAMME, H.A., CHEN, P.M., GUSTA, L.V.**: Relationship of deep supercooling and dehydration resistance to freezing injury in dormant stem tissues of 'Starkrimson Delicious' apple and 'Siberian C' peach. - J. Amer. Soc. hort. Sci. 107: 299-304, 1982.

*12367 - **QUARRIE, S.A.**: Genetic variability and heritability of drought-induced abscisic acid accumulation in spring wheat. - Plant Cell Environ. 4: 147-151, 1981.

12368 - **QUARRIE, S.A.**: Genetic differences in abscisic acid physiology and their potential uses in agriculture. - In: ADDICOTT, F.T. (ed.): Abscisic Acid. Pp. 365-419. Praeger Publishers, New York 1983.

12369 - **QUARRIE, S.A.**: Characterization of spring wheat genotypes differing in drought-induced abscisic acid accumulation. II. Leaf water relations. - J. exp. Bot. 34: 1528-1540, 1983.

12370 - **QUARRIE, S.A., LISTER, P.G.**: Characterization of spring wheat genotypes differing in drought-induced abscisic acid accumulation. I. Drought-stressed abscisic acid production. - J. exp. Bot. 34: 1260-1270, 1983.

12371 - **QUEIROZ, O.**: An hypothesis on the role of photoperiodism in the metabolic adaptation to drought. - Physiol. vég. 21: 577-588, 1983.

12372 - **QUEIROZ, O.**: Interactions between external and internal factors affecting the operation of phosphoenolpyruvate carboxylase. - Physiol. vég. 21: 963-975, 1983.

*12373 - **QUEIROZ, O., BRULFERT, J.**: Photoperiod-controlled induction and enhancement of seasonal adaptation to drought. - In: TING, I.P., GIBBS, M. (ed.): Crassulacean Acid Metabolism. Pp. 208-230. Waverly Press, Baltimore 1982.

12374 - **QUINTEIRO, M.J., GONZALES, S.M., CALZADA, C., CASTILLO, M.A., PEÑA, M.**: Efecto de la inoculation de frijol en zonas temporal en Durango. [Bean inoculation effect in areas under water stress in the state of Durango.] - Turrialba 33: 303-309, 1983. [In Span, ab: E.]

*12375 - **QURESHI, F.A., McINTYRE, G.I.**: Apical dominance in the rhizome of *Agropyron repens*: the influence of nitrogen and humidity on the translocation of ^{14}C--labelled assimilates. - Can. J. Bot. 57: 1229-1235, 1979.

12376 - **RADCHENKO, S.S., SEMIN, V.S., BERDIN, V.S., IL'NITSKIĬ, O.A.**: Opredelenie skorosti perenosa vody po ksileme rasteniĭ. [Determination of the rate of water transport along the xylem of plants.] - Fiziol. Biokhim. kul't. Rast. 15: 291-295, 1983. [In R, ab: E.]

12377 - **RADFORD, B.J., NIELSEN, R.G.H.**: Extension of crop sowing time during dry weather by means of stubble mulching and water injection. - Aust. J. exp. Agr. anim. Husb. 23: 302-308, 1983.

12378 - **RADIN, J.W.**: Physiological consequences of cellular water deficits: osmotic adjustment. - In: TAYLOR, H.M., JORDAN, W.R., SINCLAIR, T.R. (ed.): Limitations to Efficient Water Use in Crop Production. Pp. 267-276. Amer. Soc. Agron., Inc., Crop Sci. Soc. Amer., Inc., Soil Sci. Soc. Amer., Inc., Madison 1983.

*12379 - **RAGHAVULU, P., SINGH, S.P.**: Effect of mulches and transpiration suppressants on yield, water use efficiency and uptake of nitrogen and phosphorus by sorghum under dryland conditions of north-western India. - J. agr. Sci. 98: 103-108, 1982.

12380 - **RAJAGOPAL, V.**: Effect of irradiance on free proline accumulation in stressed barley leaves. - Z. Pflanzenphysiol. 111: 277-281, 1983.

12381 - **RAJENDRUDU, G., SINGH, M., WILLIAMS, J.H.**: Hydraulic press measurements of leaf water potential in groundnuts. - Exp. Agr. 19: 287-291, 1983.

12382 - **RAMOS, C., HALL, A.E.**: Effects of photon fluence rate and intercellular CO_2 partial pressure on leaf conductance and CO_2 uptake rate in *Capsicum* and *Amaranthus*. - Photosynthetica 17: 34-42, 1983.

12383 - **RAND, R.H.**: Fluid mechanics of green plants. - Annu. Rev. Fluid Mechanics 15: 29-45, 1983.

*12384 - **RAO, G.G., RAO, K.V.R., RAO, G.R.**: Studies on salt tolerance of pigeon pea cultivars. 1. Germination, seedling growth and some physiological changes. - Proc. Ind. Acad. Sci., Plant Sci. 90: 555-559, 1981.

12385 - **RAO, I.M., AMUNDSON, R.G., ALSCHER-HERMAN, R., ANDERSON, L.E.**: Effects of SO_2 on stomatal metabolism in *Pisum sativum* L. - Plant Physiol. 72: 573-577, 1983.

12386 - **RAO, I.M., ANDERSON, L.E.**: Light and stomatal metabolism. I. Possible involvement of light modulation of enzymes in stomatal movement. - Plant Physiol. 71: 451-455, 1983.

12387 - **RAO, I.M., ANDERSON, L.E.:** Light and stomatal metabolism. II. Effects of sulfite and arsenite on stomatal opening and light modulation of enzymes in epidermis. - Plant Physiol. 71: 456-459, 1983.

*12388 - **RAO, I.V.R., MOHAN RAM, H.Y.:** Effect of water stress and sucrose on opening and longevity of flowers in gladiolus. - Proc. Ind. Acad. Sci., Plant Sci. 91: 371-378, 1982.

*12389 - **RAO, N.V., INAMDAR, J.A.:** Structure and development of normal and abnormal stomata in the seedlings of some *Cruciferae*. - Proc. Ind. Acad. Sci., Plant Sci. 90: 521-533, 1981.

12390 - **RAO, N.V., INAMDAR, J.A.:** Leaf architectural studies in the *Brassicaceae*. - Bot. Mag. (Tokyo) 96: 15-28, 1983.

*12391 - **RAO, S.R.S., RAMAYYA, N.:** Distribution of stomata and its relation to plant habit in the order *Malvales*. - Ind. J. Bot. 4: 149-156, 1981.

12392 - **RAPER, C.D., Jr., KRAMER, P.J. (ed.):** Crop Reactions to Water and Temperature Stresses in Humid, Temperate Climates. - Westview Press, Boulder 1983.

12393 - **RASMUSSEN, H.:** Stomatal development in families of *Liliales*. - Bot. Jahrb. Syst. Pflanzengesch. Pflanzengeogr. 104: 261-287, 1983.

12394 - **RASULOV, B.Kh.:** O vnutrennikh faktorakh, ogranichivayushchikh vidimyĭ fotosintez v ontogeneze lista khlopchatnika. [Internal factors limiting apparent photosynthesis during cotton leaf ontogenesis.] - Dokl. Akad. Nauk Tadzh. SSR 24: 573-576, 1981. [In R.]

*12395 - **RASULOV, B.Kh.:** Reassimilyatsiya uglekisloty, obrazovannoĭ v rezul'tate dykhatel'nykh protsessov na svetu, v ontogeneze lista u dvukh vidov khlopchatnika. [Reassimilation of CO_2, the product of respiration processes in the light, during leaf ontogeny in two cotton species.] - Dokl. Akad. Nauk Tadzh. SSR 25: 250-254, 1982. [In R.]

12396 - **RASULOV, B.Kh., LAĬSK, A.Kh., ASROROV, K.A.:** Izmeneniya pokazateleĭ fotosinteza v ontogeneze lista u dvukh vidov khlopchatnika. [Photosynthetic parameters during leaf ontogeny in two cotton species.] - Fiziol. Rast. 30: 305-314, 1983. [In R, ab: E.]

12397 - **RASZKA, P.:** Tvorba výnosu ozimé pšenice seté do nezpracované půdy a zavlažované do zásoby. [Yield formation in winter wheat directly sown and with store irrigation.] - Rost. Výroba (Praha) 29: 817-826, 1983. [In Czech, ab: R, E.]

*12398 - **RATHERT, G.:** Influence of extreme K:Na ratios and high substrate salinity on plant metabolism of crops differing in salt tolerance. VII. Relations between carbohydrates and degradative enzymes in salt tolerant and salt sensitive cotton genotypes during initial salinity stress. - J. Plant Nutr. 5: 1401-1413, 1982.

12399 - **RATHERT, G.:** Effects of high salinity stress on mineral and carbohydrate metabolism of two cotton varieties. - Plant Soil 73: 247-256, 1983.

12400 - **RAVINDRANATH, P., RAJ, A.S.:** Effect of moisture stress on growth, yield and yield components of field-grown sorghum varieties having glossy and non--glossy leaves. - Ind. J. agr. Sci. 53: 428-430, 1983.

*12401 - **RAWSON, H.M., CONSTABLE, G.A.:** Gas exchange of pigeonpea: a comparison with other crops and a model of carbon production and its distribution within the plant. - In: International Workshop on Pigeonpeas. Volume 1. Pp. 175-189. ICRISAT Center, Patancheru (Andhra Pradesh) 1980.

12402 - **RAWSON, H.M., HINDMARSH, J.H.:** Light, leaf expansion and seed yield in sunflower. - Aust. J. Plant Physiol. 10: 25-30, 1983.

12403 - **RAWSON, H.M., TURNER, N.C.:** Irrigation timing and relationships between leaf area and yield in sunflower. - Irrig. Sci. 4: 167-175, 1983.

12404 - **RAYDER, L., TING, I.P.:** CAM-idling in *Hoya carnosa* (*Asclepiadaceae*). - Photosynthesis Res. 4: 203-211, 1983.

12405 - **RAYDER, L., TING, I.P.:** Shifts in the carbon metabolism of *Xerosicyos danguyi* H. Humb. (*Cucurbitaceae*) brought about by water stress. I. General characteristics. - Plant Physiol. 72: 606-610, 1983.

12406 - **RAYDER, L., TING, I.P.:** Shifts in the carbon metabolism of *Xerosicyos danguyi* H. Humb. (*Cucurbitaceae*) brought about by water stress. II. Enzymology. - Plant Physiol. 72: 611-615, 1983.

*12407 - **RAYLE, D.L., ROSS, C.W., ROBINSON, N.:** Estimation of osmotic parameters accompanying zeatin-induced growth of detached cucumber cotyledons. - Plant Physiol. 70: 1634-1636, 1982.

*12408 - **REDDY, K.A., VENKATACHARI, A.:** Consumptive use of water by wheat in Andhra Pradesh. - Ind. J. agr. Res. 16: 104-108, 1982.

*12409 - **REDDY, K.A., VENKATACHARI, A., RAO, P.S.P.:** Evapotranspiration and water-use efficiency of different crops. - Ind. J. Agron. 25: 176-180, 1980.

12410 - **REDDY, K.B., RAO, K.R., RAGHAVENDRA, A.S.:** Stomatal metabolism in isolated abaxial and adaxial epidermis of *Commelina benghalensis*: variation in proton efflux and adenosine triphosphatase activity. - Z. Pflanzenphysiol. 111: 401 - 412, 1983.

12411 - **REDDY, K.R., TUCKER, J.C.:** Productivity and nutrient uptake of water hyacinth, *Eichhornia crassipes*. I. Effect of nitrogen source. - Econ. Bot. 37: 237-247, 1983.

*12412 - **REDDY, M.D., KRISHNA MURTHY, I., REDDY, K.A., VENKATACHARI, A.:** Consumptive use and daily evapotranspiration of corn under different levels of nitrogen and moisture regimes. - Plant Soil. 56: 143-147, 1980.

12413 - **REDDY, M.G., RAO, Y.Y., RAO, K.S., RAMASESHAIAH, K.:** A preliminary study on scheduling irrigation with can evaporimeters. - Agr. Water Manage. 6: 403-407, 1983.

12414 - **REDDY, S.J.:** Irrigation scheduling using ICSWAB model. - Pesq. agropec. Bras. 18: 1121-1126, 1983.

12415 - **REDON, J., LANGE, O.L.:** Epiphytische Flechten im Bereich einer chilenischen "Nebeloase" (Fray Jorge). I. Vegetationskundliche Gliederung und Standortsbedingungen. - Flora 174: 213-243, 1983.

12416 - **REED, R.H.:** Measurement and osmotic significance of β-dimethylsulphoniopropionate in marine macroalgae. - Mar. Biol. Lett. 4: 173-181, 1983.

12417 - **REED, R.H., BARRON, J.A.:** Physiological adaptation to salinity change in *Pilayella littoralis* from marine and estuarine sites. - Bot. mar. 26: 409-416, 1983.

12418 - **REED, R.H., STEWART, W.D.P.:** Physiological responses of *Rivularia atra* to salinity: osmotic adjustment in hyposaline media. - New Phytol. 95: 595-603, 1983.

12419 - **REICOSKY, D.C.:** Soil management for efficient water use: soil-profile modification effects on plant growth and yield in the southwestern United States. - In: TAYLOR, H.M., JORDAN, W.R., SINCLAIR, T.R. (ed.): Limitations to Efficient Water Use in Crop Production. Pp. 471-477. Amer. Soc. Agron., Inc., Crop Sci. Soc. Amer., Inc., Soil Sci. Soc. Amer., Inc., Madison 1983.

12420 - **REICOSKY, D.C., SHARRATT, B.S., LJUNGKULL, J.E., BAKER, D.G.:** Comparison of alfalfa evapotranspiration measured by a weighing lysimeter and a portable chamber. - Agr. Meteorol. 28: 205-211, 1983.

12421 - **RENARD, C., DEMESSEMACKER, W.:** Effects of wind velocity on stomatal conductance and consequences of leaf rolling on water uptake in tall fescue. - Biol. Plant. 25: 408-411, 1983.

12422 - **RENGER, M., STREBEL, O.:** Einfluss von Grundwassersenkungen auf den Pflanzenertrag bei Acker- und Grünland. - Kali-Briefe 16: 379-389, 1983.

12423 - **RENGER, M., STREBEL, O.:** Einfluss des Grundwasserflurabstandes auf Grundwasserneubildung, Evapotranspiration und Pflanzenertrag. - Z. Deut. geol. Ges. 134: 669-678, 1983.

*12424 - **RENVOIZE, S.A.:** A survey of leaf-blade anatomy in grasses. I. *Andropogoneae*. - Kew Bull. 37: 315-321, 1982.

*12425 - **RENVOIZE, S.A.:** A survey of leaf-blade anatomy in grasses. II. *Arundinelleae*. - Kew Bull. 37: 489-495, 1982.

*12426 - **RENVOIZE, S.A.:** A survey of leaf-blade anatomy in grasses. III. *Garnotieae*. - Kew Bull. 37: 497-500, 1982.

12427 - **RENVOIZE, S.A.:** A survey of leaf-blade anatomy in grasses. IV. *Eragrostideae*. - Kew Bull. 38: 469-478, 1983.

12428 - **RIBEIRO, A.S., SILVEIRA, P., Jr., TERRES, A.L.S.:** Épocas de semeadura em arroz irrigado. [Seeding time on irrigated rice.] - Pesq. agropec. Bras. 18: 619-630, 1983.[In Port, ab: E.]

12429 - **RICHARDS, R.A.:** Manipulation of leaf area and its effect on grain yield in droughted wheat. - Aust. J. agr. Res. 34: 23-31, 1983.

12430 - **RICHTER, H., WAGNER, S.B.:** Water stress resistance of photosynthesis: some aspects of osmotic relations. - In: MARCELLE, R., CLIJSTERS, H., VAN POUCKE, M. (ed.): Effects of Stress on Photosynthesis. Pp. 45-53. Martinus Nijhoff / Dr W. Junk Publishers, The Hague - Boston - London 1983.

*12431 - **RICHTER, W.:** Zur Abhängigkeit des Winterweizenertrages auf einem anlehmigen Sandboden vom Niederschlagsangebot im Hauptwasserbedarfzeitraum. - Arch. Acker- Pflanzenbau Bodenk. 26: 553-559, 1982.

*12432 - **RICHTER, W., SCHÄFER, W.:** Untersuchungen zur Einfluss der Beregnung auf die Kartoffelerträge auf leichten Diluvialstandorten. - Arch. Acker- Pflanzenbau Bodenk. 24: 461-466, 1980.

*12433 - **RICKLEFS, R.E., MATTHEW, K.K.:** Chemical characteristics of the foliage of some deciduous trees in southeastern Ontario. - Can. J. Bot. 60: 2037-2045, 1982.

12434 - **RIEHL, T.E., UNGAR, I.A.:** Growth, water potential and ion accumulation in the inland halophyte *Atriplex triangularis* under saline field conditions. - Acta oecol. - Oecol. Plant. 4: 27-39, 1983.

12435 - **RITCHIE, J.T.:** Reduction of stresses related to soil water deficit. - In: RAPER, C.D., Jr., KRAMER, P.J. (ed.): Crop Reactions to Water and Temperature Stresses in Humid, Temperate Climates. Pp. 329-340. Westview Press, Boulder 1983.

12436 - **RITCHIE, J.T.:** Efficient water use in crop production: discussion on the generality of relations between biomass production and evapotranspiration. - In: TAYLOR, H.M., JORDAN, W.R., SINCLAIR, T.R. (ed.): Limitations to Efficient Water Use in Crop Production. Pp. 29-44. Amer. Soc. Agron., Inc., Crop Sci. Soc. Amer., Inc., Soil Sci. Soc. Amer., Inc., Madison 1983.

12437 - **ROBBERECHT, R., MAHALL, B.E., NOBEL, P.S.:** Experimental removal of intraspecific competitors - effects on water relations and productivity of a desert bunchgrass, *Hilaria rigida*. - Oecologia 60: 21-24, 1983.

12438 - **ROBERTS, J.:** Forest transpiration: a conservative hydrological process? - J. Hydrol 66: 133-141, 1983.

12439 - **ROBERTSON, B.L.:** Leaf develópment in *Jubaeopsis caffra* (Becc.). - Ann. Bot. 51: 555-557, 1983.

12440 - **ROBINSON, M.:** Influence of abscisic acid and ethylene on assimilate distribution in *Gladiolus grandiflorus*. - Ann. Bot. 51: 779-785, 1983.

12441 - **ROBINSON, M., HALEVY, A.H., GALILI, D., PLAUT, Z.:** Distribution of assimilates in *Gladiolus grandiflorus* as affected by water deficit. - Ann. Bot. 51: 461-468, 1983.

12442 - **ROBINSON, R.G.:** Yield and composition of field bean and adzuki bean in response to irrigation, compost, and nitrogen. - Agron. J. 75: 31-35, 1983.

*12443 - **ROBINSON, R.G., FORD, J.H., LUESCHEN, W.E., RABAS, D.L., SMITH, L.J., WARNES, D.D., WIERSMA, J.V.:** Response of sunflower to plant population. - Agron. J. 72: 869-871, 1980.

*12444 - **ROBINSON, R.G., FORD, J.H., LUESCHEN, W.E., RABAS, D.L., WARNES, D.D., WIERSMA, J.V.:** Sunflower plant population and its arrangement. - Minnesota agr. Exp. Sta. Misc. Rep. 180: 1-8, 1982.

*12445 - **ROBINSON, R.G., FORD, J.H., LUESCHEN, W.E., RABAS, D.L., WARNES, D.D., WIERSMA, J.V.:** Response of sunflower to uniformity of plant spacing. - Agron. J. 74: 363-365, 1982.

12446 - **ROBINSON, S.P., DOWNTON, W.J.S., MILLHOUSE, J.A.:** Photosynthesis and ion content of leaves and isolated chloroplasts of salt-stressed spinach. - Plant Physiol. 73: 238-242, 1983.

12447 - **ROCHER, J.-P.:** Les transports de substances (gas et assimilats) dans les phases gazeuses et liquides des feuilles. - Bull. Soc. bot. France, Actualiés Bot. 130: 99-112, 1983.

12448 - **ROCKWOOD, D.L., COMER, C.W., DIPPON, D.R., HUFFMAN, J.B., RIEKERK, H., WANG, S.:** Current status of woody biomass production research in Florida. - Soil Crop Sci. Soc. Florida Proc. 42: 19-27, 1983.

12449 - **RODRIGUES, J.D., KLAR, A.E., PEDRAS, J.F., RODRIGUES, S.D., DELACHIAVE, M.E. A.:** A influência de diferentes regimes de umidade do solo no desenvolvimento da parte aérea de gladiolos. [The influence of different soil humidity levels on the development of the aerial part of the gladiolus.] - Fyton 43: 81-95, 1983. [In Port, ab: E.]

12450 - **ROEB, G.W., WIENEKE, J., FÜHR, F.:** Wirkung einer Abscisinsäure-Applikation auf die Cl^--Translokation in Sojabohnen bei niedriger und hoher Besalzung des Nährmediums. - Z. Pflanzenarnähr. Bodenk. 146: 385-397, 1983.

12451 - **ROGERS, H.H., BINGHAM, G.E., CURE, J.D., SMITH, J.M., SURANO, K.A.:** Responses of selected plant species to elevated carbon dioxide in the field. - J. environ. Qual. 12: 569-574, 1983.

12452 - **ROGERS, J.S., ALLEN, L.H., Jr., CALVERT, D.V.:** Evapotranspiration from a humid-region developing citrus grove with grass cover. - Trans. ASAE 26: 1778-1783, 1792, 1983.

12453 - **ROGERS, J.S., STANLEY, C.D.:** Subsurface irrigation of staked tomatoes in Florida. - Soil Crop Sci. Soc. Florida Proc. 42: 65-69, 1983.

12454 - **ROJAS, L., RECIO, O., De CASTRO, M.E.:** Xeromorfismo foliar y resistencia a la sequía en dos variedades de caña de azúcar. [Leaf xeromorphy and drought resistance of sugarcane varieties.] - Cienc. Agr. 16: 33-37, 1983. [In Span, ab: E.]

*12455 - **ROJIANI, K.B., ROSS, B.B., WOESTE, F.E., SHANHOLTZ, V.O.:** A probabilistic model for assessment of plant water availability. - Trans. ASAE 25: 1576-1580, 1588, 1982.

*12456 - **ROSINGER, C.H., WILSON, J.M., KERR, M.W.:** Changes in the temperature response of Hill-reaction activity of chilling-sensitive and chilling-resistant plants after hardening. - J. exp. Bot. 33: 321-331, 1982.

*12457 - **ROTH, R., ROTH, D., WEBER, W.:** Untersuchungen zum Einfluss unterschiedlicher Wasserversorgung auf den Ertrag von Silomais. - Arch. Acker- Pflanzenbau Bodenk. 26: 605-614, 1982.

12458 - **ROUGHLEY, R.J., SPRENT, J.I., DAY, J.M.:** Nitrogen fixation. - In: HEBBLETHWAITE, P.D. (ed.): The Faba Bean (*Vicia faba* L.). A Basis for Improvement. Pp. 233-269. Butterworths, London - Boston - Durban - Singapore - Sydney - Toronto - Wellington 1983.

12459 - **ROUNDY, B.A.:** Response of basin wildrye and tall wheatgrass seedlings to salination. - Agron. J. 75: 67-70, 1983.

12460 - **ROWSE, H.R., MASON, W.K., TAYLOR, H.M.:** A microcomputer simulation model of soil water extraction by soybeans. - Soil Sci. 136: 218-225, 1983.

12461 - **ROY, J., BERGER, A.:** Water potential measurement, water compartmentation and water flow in *Dactylis glomerata* L. leaves. - New Phytol. 93: 43-52, 1983.

12462 - **ROZEMA, J., Van MANEN, Y., VUGTS, H.F., LEUSINK, A.:** Airborne and soilborne salinity and the distribution of coastal and inland species of the genus *Elytrigia*. - Acta Bot. Neerl. 32: 447-456, 1983.

12463 - **RUČKA, M.:** Vplyv závlahy a hnojenia na kvalitatívne ukazatele ozimnej pšenice. [The effect of irrigation and fertilization on the quality parameters of winter wheat.] - Rost. Výroba (Praha) 29: 875-884, 1983. [In Slov, ab: R, E.]

12464 - **RUČKA, M.:** Vplyv stupňovaných dávok dusíka, fosforu a foriem dusíkatých hnojív pri rôznych vláhových režimoch na produkciu krmných hodnôt kukurice na siláž. [The effect of increasing nitrogen and phosphorus rates and forms of nitrogen fertilizers on the feeding value of silage maize grown in different moisture regimes.] - Rost. Výroba (Praha) 29: 1233-1243, 1983. [In Slov, ab: R, E.]

*12465 - **RUDICH, J.:** Growing of processing-tomato plants under water-deficiency conditions: mulching with transparent polyethylene. - Scientia Hort. 10: 117-125, 1979.

*12466 - **RUDRAPAL, A.B., BASU, R.N.:** Carbohydrate utilization in wheat seed in relation to viability maintenance by soaking-drying pretreatments. - Seed Res. 7: 5-10, 1979.

*12467 - **RUGGIERO, C.:** Roots distribution of tomato plants with relation to soil water regimen and to irrigation method. - In: Seminares sur l'Irrigation Localisée. III. Influence de l'Irrigation Localisée sur la Morphologie et la Physiologie des Racines. (Pubbl. N. 49.) Pp. 145-154. Consiglio Nazionale delle Ricerche, Napoli - Ponticelli, 1980.

*12468 - **RUGGIERO, C., XILOYANNIS, C., URIU, K.:** Variations in water use as indication of root distribution in irrigated by flood, drip and non-irrigated peach trees. - In: Seminares sur l'Irrigation Localisée. III. Influence de l'Irrigation Localisée sur la Morphologie et la Physiologie des Racines. (Pubbl. N. 50.) Pp. 103-114. Consiglio Nazionale delle Ricerche, Napoli - Ponticelli 1980.

*12469 - **RUGGIERO, C., XILOYANNIS, C., URIU, K., BEUTEL, J., FERRERES, E.:** Water consumptive use in irrigated by flood, drip and non-irrigated peach trees. - Acta Hort. 119: 159-178, 1981.

12470 - **RÜHL, E., ALLEWELDT, G.:** Untersuchungen zum Gaswechsel der Rebe. III. Einfluss vorübergehender Trockenheit auf die Leistungsfähigkeit verschiedener Rebsorten. - Vitis 22: 120-128, 1983.

*12471 - **RÜHL, E.-H.:** Einfluss der Wasserversorgung auf Photosynthese, Transpiration und vegetatives Wachstum verschiedener Rebsorten. - Thesis, Fakultät III - Agrarwissenschaften I, Universität Hohenheim, Stuttgart-Hohenheim 1981.

12472 - **RUMBAUGH, M.D., JOHNSON, D.A., RINEHART, D.N.:** Stand density, shoot weight, and acetylene reduction activity of alfalfa populations subjected to field and greenhouse moisture gradients. - Crop Sci. 23: 784-789, 1983.

12473 - **RUNGE, M.:** Physiology and ecology of nitrogen nutrition. - In: LANGE, O.L., NOBEL, P.S., OSMOND, C.B., ZIEGLER, H. (ed.): Physiological Plant Ecology III. Responses to the Chemical and Biological Environment. Pp. 163-200. Springer-Verlag, Berlin - Heidelberg - New York 1983.

12474 - **RUSSO, D.:** Crop yield - irrigation relationships in a gypsiferous-sodic soil. - Agron. J. 75: 427-434, 1983.

*12475 - **RYCHNOVSKÁ, M.:** Vodní provoz přirozených a rekultivovaných porostů v podhorské oblasti. [Water regime of natural and recultivated stands in hilly region.] - Acta Univ. Agr., Fac. Agron. (Brno) A28 (3-4): 285-288, 1980. [In Czech.]

12476 - **RYGOL, J., LÜTTGE, U.:** Water-relation parameters of giant and normal cells of *Capsicum annuum* pericarp. - Plant Cell Environ. 6: 545-553, 1983.

*12477 - **SABALE, A.B.:** Note on organic and inorganic constituents in *Cymbopogon martnii* (Roxb.) Wats. during different stages of growth. - Ind. J. agr. Sci. 52: 199-201, 1982.

12478 - **SACHER, J.A.:** Abscisic acid in leaf senescence. - In: ADDICOTT, F.T. (ed.): Abscisic Acid. Pp. 479-522. Praeger Publishers, New York 1983.

12479 - **SACHER, R.F., STAPLES, R.C., ROBINSON, R.W.:** Ion regulation and response of tomato to sodium chloride: a homeostatic system. - J. Amer. Soc. hort. Sci. 108: 566-569, 1983.

12480 - **SACK, F., PAOLILLO, D.J., Jr.:** Structure and development of walls in *Funaria* stomata. - Amer. J. Bot. 70: 1019-1030, 1983.

12481 - **SACK, F.D., PAOLILLO, D.J., Jr.:** Protoplasmic changes during stomatal development in *Funaria*. - Can. J. Bot. 61: 2515-2526, 1983.

12482 - **SACK, F.D., PAOLILLO, D.J., Jr.:** Stomatal pore and cuticle formation in *Funaria*. - Protoplasma 116: 1-13, 1983.

*12483 - **SAHA, R., BASU, R.N.:** Maintenance of soybean seed viability by hydration-dehydration treatments. - Ind. Agr. 25: 275-278, 1981.

*12484 - **SAHU, T.R., MISHRA, D.P.:** A new method for stomatal studies. - Comp. Physiol. Ecol. 6: 280, 1981.

12485 - **SAILEROVÁ, E.:** Vliv teploty a vodního sytostního deficitu na kompenzační koncentraci CO_2 u listu mladých rostlin ječmene. [Effect of temperature and water saturation deficit on CO_2 compensation concentration in leaves of young barley plants.] - In: Dny Rostlinné Fyziologie III. Pp. 280-281. ČSBS, Praha 1983. [In Czech.]

12486 - **SAINI, B.C., RATHORE, T.R.:** Adaptation of beta-ray gauge technique for measurement of water deficit of intact rice leaves. - Biol. Plant. 25: 326-330, 1983.

12487 - **SAINI, B.C., RATHORE, T.R.:** Diurnal changes in water deficit and recovery patterns of intact rice leaves under natural and controlled environment conditions. - Biol. Plant. 25: 331-335, 1983.

*12488 - **SALA, O.E., LAUENROTH, W.K., REID, C.P.P.:** Water relations: A new dimension for niche separation between *Bouteloua gracilis* and *Agropyron smithii* in North American semi-arid grasslands. - J. appl. Ecol. 19: 647-657, 1982.

12489 - **SALASOO, I.:** Effect of leaf age on epicuticular wax alkanes in *Rhododendron*. - Phytochemistry 22: 461-463, 1983.

12490 - **SALCHEVA, G., GEORGIEVA, D.:** Vliyanie na pochvenata vlazhnost v"rkhu s"d"rzhanieto na azot, fosfor i kaliĭ v pshenichni rasteniya prez zimno-proletniya period. [The effect of soil moisture on nitrogen, phosphorus and potassium content in wheat plants during the winter-spring period.] - Fiziol. Rast. (Sofia) 9 (2): 42-51, 1983. [In Bulg, ab: R, E.]

12491 - **SALIM, M., PITMAN, M.G.:** Effects of salinity on ion uptake and growth of mung bean plants (*Vigna radiata* L.). - Aust. J. Plant Physiol. 10: 395-407, 1983.

12492 - **SALLEO, S.:** Water relations parameters of two Sicilian species of *Senecio* (Groundsel) measured by the pressure bomb technique. - New Phytol. 95: 179-188, 1983.

12493 - **SALOMONSSON, A.-C., BERTHOLDSSON, N.-O., STOY, V., THEANDER, O., ÅMAN, P.:** Carbohydrate composition of and ^{14}C incorporation into developing grains of normal and high-lysine genotypes of barley. - Swed. J. agr. Res. 13: 75-89, 1983.

12494 - **SAMARIN, N.A., SAMARINA, L.N.:** Produktivnost' gorokha ovoshchnogo v zavisimosti ot vodoobespecheniya. [Produktivity of garden pea as related to water supply.] - Sel'skokhoz. Biol. 1983 (4): 9-12, 1983. [In R, ab: E.]

*12495 - **SAMIEV, Kh.S., MARFINA, K.G.:** Ëlektroforeticheskiĭ analiz belkov khloroplastov kak kriteriĭ ustoĭchivosti khlopchatnika k zasukhe. [Electrophoretic analysis of chloroplast proteins as an index for drought resistance of cotton plants.] - Dokl. Vses. Akad. sel'skokhoz. Nauk Im. V.I. Lenina 1982 (6): 19-21, 1982. [In R.]

*12496 - **SANCES, F.V., TOSCANO, N.C., JOHNSON, M.W., LaPRÉ, L.F.:** Pesticides may reduce lettuce yield. - Calif. Agr. 35 (11/12): 4-5, 1981.

12497 - **SÁNCHEZ, R.A., De MIGUEL, L.C.:** Ageing of *Datura ferox* seeds embryos during dry storage and its reversal during imbibition. - Z. Pflanzenphysiol. 110: 319-329, 1983.

12498 - **SÁNCHEZ, R.A., HALL, A.J., TRAPANI, N., COHEN de HUNAU, R.:** Effects of water stress on the chlorophyll content, nitrogen level and photosynthesis of leaves of two maize genotypes. - Photosynthesis Res. 4: 35-47, 1983.

12499 - **SANDERSON, J.:** Water uptake by different regions of the barley root. Pathways of radial flow in relation to development of the endodermis. - J. exp. Bot. 34: 240-253, 1983.

*12500 - **SANDHU, B.S., KHERA, K.L., PRIHAR, S.S., SINGH, B.:** Irrigation needs and yield of rice on a sandy-loam soil as affected by continuous and intermittent submergence. - Ind. J. agr. Sci. 50: 492-496, 1980.

*12501 - **SANDHU, B.S., KHERA, K.L., SINGH, B.:** Note on the use of irrigation water and yield of transplanted rice in relation to timing of last irrigation. - Ind. J. agr. Sci. 52: 870-871, 1982.

*12502 - **SANDHU, B.S., PRIHAR, S.S., KHERA, K.L.:** Sugarcane response to irrigation and straw mulch in a subtropical region. - Agr. Water Manage. 3: 35-44, 1980.

12503 - **SANDSBORG, J., RODSKJER, N.:** Changes of soil water content under winter wheat, barley and bare soil during the growing season. A field study. - Swed. J. agr. Res. 13: 17-29, 1983.

12504 - **SAN JOSÉ, J.J.:** Diurnal course of CO_2 and water vapour exchange in *Manihot esculenta* Crantz var. cubana. - Photosynthetica 17: 12-19, 1983.

12505 - **SAN JOSÉ, J.J., BERRADE, F.:** Transfer of mass and energy in a cassava (*Manihot esculenta* Crantz cv. Cubana) community 1. Microclimate and water vapour exchange in a savanna climate. - Ann. Bot. 52: 507-520, 1983.

12506 - **SAN JOSÉ, J.J., CABRERA, M.:** Energy and gas exchange within a sunflower (*Helianthus annuus* L. cv. Manfredi) canopy. - Photosynthetica 17: 412-421, 1983.

12507 - **SANT, F.I.:** A mechanical viewing aid for microscopical examination of convoluted leaf surfaces of forage grasses. - J. exp. Bot. 34: 765-768, 1983.

12508 - **SANTARIUS, K.A., GIERSCH, C.:** Inactivation and protection of isolated thylakoid membranes during freezing. - In: MARCELLE, R., CLIJSTERS, H., VAN POUCKE, M. (ed.): Effects of Stress on Photosynthesis. Pp. 201-210. Martinus Nijhoff / Dr W. Junk Publishers, The Hague - Boston - London 1983.

12509 - **SARIĆ, M.R., LOUGHMAN, B.C. (ed.):** Genetic Aspects of Plant Nutrition. (Developments in Plant and Soil Sciences Volume 8. Martinus Nijhoff Publishers, The Hague - Boston - Lançaster 1983.

12510 - **SATLER, S.O., THIMANN, K.V.:** Metabolism of oat leaves during senescence. VII. The interaction of carbon dioxide and other atmospheric gases with light in controlling chlorophyll loss and senescence. - Plant Physiol. 71: 67-70, 1983.

12511 - **SATOH, K., SMITH, C.M., FORK, D.C.:** Effects of salinity on primary processes of photosynthesis in the red alga *Porphyra perforata*. - Plant Physiol. 73: 643-647, 1983.

12512 - **SAUTER, J.J.:** Efflux and reabsorption of sugars in the xylem. II. Seasonal changes in sucrose uptake in *Salix*. - Z. Pflanzenphysiol. 111: 429-440, 1983.

12513 - **SAVAGE, M.J., WIEBE, H.H., CASS, A.:** *In situ* field measurement of leaf water potential using thermocouple psychrometers. - Plant Physiol. 73: 609-613, 1983.

12514 - **SAXE, H.:** Long-term effects of low levels of SO_2 on bean plants (*Phaseolus vulgaris*). I. Immission-response pattern of net photosynthesis and transpiration during life-long, continuous measurements. - Physiol. Plant. 57: 101-107, 1983.

*12515 - **SAXTON, K.E.:** Mathematical modeling of evapotranspiration on agricultural watersheds. - In: Modeling Components of Hydrologic Cycle. Pp. 183-204. Water Resources Publications, Littleton 1982.

*12516 - **SCALA, A., QUAGLIETTA CHIARANDÀ, F.:** Effect of water potential on *Rhizoctonia* damping-off of cauliflower. - In: Proceedings of the Fifth Congress of Mediterranean Phytopathological Union. (Pubbl. N. 48.) Pp. 132-135. Consiglio Nazionale delle Ricerche, Napoli - Ponticelli 1980.

12517 - **SCHÄFER, W., KLANK, I., WEIRAUCH, M., REIHER, W.:** System zur automatischen Erfassung von Messwerten der CO_2-Assimilation und Transpiration unter Feldbedingungen. 2. Mitteilung: Messprogramm und Programmsystem. - Arch. Acker-Pflanzenbau Bodenk. 27: 361-366, 1983.

12518 - **SCHLEIFF, U.:** Salztoleranz von Mais bei Applikation salzreichen Giesswassers in unterschiedlichen Intervallen. - Z. Acker- Pflanzenbau 152: 9-18, 1983.

12519 - **SCHLEIFF, U.:** Water uptake of barley roots from rhizospheric soil solutions of different salt concentrations. - Irrig. Sci. 177-189, 1983.

12520 - **SCHLICHTER, T.M., VAN DER PLOEG, R.R., ULRICH, B.:** A simulation model of the water uptake of a beech forest: resting variations in root biomass and distribution. - Z. Pflanzenernähr. Bodenk. 146: 725-735, 1983.

12521 - **SCHMIDT, A., SAMMLER, P., SABEH, G., BERGMANN, H.:** Retention generativer Organe von *Vicia faba* L. durch Behandlung mit $AgNO_3$ unter Wasserstressbedingungen. - Arch. Acker- Pflanzenbau Bodenk. 27: 287-295, 1983.

12522 - **SCHMIDT, J.W.:** Drought resistance and wheat breeding. - Agr. Water Manage. 7: 181-194, 1983. Also in: STONE, J.F., WILLIS, W.O. (ed.): Plant Production and Management under Drought Conditions. Pp. 181-194. Elsevier Scientific Publishing Company, Amsterdam 1983.

12523 - **SCHMIDT, W.:** Ecosystem research (Ecological geobotany). - Progress Bot. 45: 351-389, 1983.

12524 - **SCHNABL, H.:** The key role of phosphoenolpyruvate carboxylase during the volume changes of guard cell protoplasts. - Physiol. vég. 21: 955-962, 1983.

12525 - **SCHOENEWEISS, D.F.:** Drought predisposition to Cytospora canker in blue spruce. - Plant. Dis. 67: 383-385, 1983.

*12526 - **SCHONBECK, M.W., EGLEY, G.H.:** Redroot pigweed (*Amaranthus retroflexus*) seed germination responses to afterripening, temperature, ethylene, and some other environmental factors. - Weed Sci. 28: 543-548, 1980.

12527 - **SCHULTE, P.J., MARSHALL, P.E.:** Growth and water relations of black locust and pine seedlings exposed to controlled water stress. - Can. J. Forest Res. 13: 334-338, 1983.

12528 - **SCHULZE, E.-D.:** Photosynthetic CO_2 uptake and whole plant growth as related to plant water relations. - Ber. Deut. bot. Ges. 96: 391-402, 1983.

*12529 - **SCHULZE, E.-D.:** Photosynthetic production and water relations in different plant-forms as related to altitudinal gradients. - New Zeal. Forest Serv., Forest Res. Inst. Tech. Paper 70: 213-229, 1980.

12530 - **SCHULZE, E.-D., SCHILLING, K., NAGARAJAH, S.:** Carbohydrate partitioning in relation to whole plant production and water use of *Vigna unguiculata* (L.) Walp. - Oecologia 58: 169-177, 1983.

12531 - **SCHWAB, K.B., HEBER, U.:** Photosynthesis and drought - a study involving resurrection plants. - In: Sixth International Congress on Photosynthesis. Abstracts. Volume 1. Pp. 401. Brussels 1983.

12332 - **SCHWARZ, M., GALE, J.:** The effect of heath and salinity stress on the carbon balance of *Xanthium strumarium*. - In: MARCELLE, R., CLIJSTERS, H., VAN POUCKE, M. (ed.): Effects of Stress on Photosynthesis. Pp. 325-331. Martinus Nijhoff / Dr W. Junk Publishers, The Hague - Boston - London 1983.

12533 - **SCHWINTZER, C.R., LANCELLE, S.A.:** Effect of water-table depth on shoot growth, root growth, and nodulation of *Myrica gale* seedlings. - J. Ecol. 71: 489-501, 1983.

12534 - **SCIFRES, C.J., MUTZ, J.L., WHITSON, R.E., DRAWE, D.L.:** Mixed-brush canopy cover-rainfall interrelationships with native grass production. - Weed Sci. 31: 1-4, 1983.

12535 - **SECOR, J., SHIBLES, R., STEWART, C.R.:** Metabolic changes in senescing soybean leaves of similar plant ontogeny. - Crop Sci. 23: 106-110, 1983.

12536 - **SELIM, H.M., MELVILLE, D.R., KINGERY, W.L.:** Scheduling supplemental irrigation for cotton. - Louisiana Agr. 26: 12-14, 1983.

*12537 - **SEN, D.N.:** Environment and Plant Life in Indian Desert. - Geobios International, Jodhpur 1982.

*12538 - **SEN, D.N., RAJPUROHIT, K.S. (ed.):** Contributions to the Ecology of Halophytes. - Dr W. Junk Publishers, The Hague - Boston - London 1982.

*12539 - **SEN, D.N., RAJPUROHIT, K.S., WISSING, F.W.:** Survey and adaptive biology of halophytes in the western Rajasthan, India. - In: SEN, G.V., RAJPUROHIT, K. S. (ed.): Contributions to the Ecology of Halophytes. Pp. 61-78. Dr W. Junk Publishers, The Hague - Boston - London 1982.

*12540 - **SEN, H.S., SURIYA RAO, A.V., MAHATA, K.R., DAS, P.C.:** Water diffusivity of germinating soya-bean, cow-pea and blackgram seeds at both large and small times. - J. agr. Sci. 98: 209-213, 1982.

12541 - **SEN, U.:** Stomatal structure and stomatogenesis in *Azolla pinnata* R. Brown. - Ann. Bot. 52: 201-204, 1983.

12542 - **SENARATNA, T., McKERSIE, B.D.:** Dehydration injury in germinating soybean (*Glycine max* L. Merr.) seeds. - Plant Physiol. 72: 620-624, 1983.

12543 - **SENARATNA, T., McKERSIE, B.D.:** Characterization of solute efflux from dehydration injured soybean (*Glycine max* L. Merr) seeds. - Plant Physiol. 72: 911-914, 1983.

12544 - **ŠESTÁK, Z., BENEŠOVÁ, H., ZIMA, J., POSPÍŠILOVÁ, J., KUTÍK, J.:** Effects of age and water potential of leaves on photochemical activities of immobilized chloroplasts. - In: Sixth International Congress on Photosynthesis. Abstracts, Volume 1. Pp. 403. Brussels 1983.

12545 - **SETTER, T.L., FLANNIGAN, B.A.:** Time course of photosynthesis and stomatal conductance following changes in light flux density. - Crop Sci. 23: 795-797, 1983.

12546 - **SETTER, T.L., GREENWAY, H.:** Changes in the proportion of endogenous osmotic solutes accumulated by *Chlorella emersonii* in the light and dark. - Plant Cell Environ. 6: 227-234, 1983.

12547 - **SHACKEL, K.A., HALL, A.E.:** Comparison of water relations and osmotic adjustment in sorghum and cowpea under field conditions. - Aust. J. Plant Physiol. 10: 423-435, 1983.

12548 - **SHAH, J.J., NAIR, G.M., THANKI, Y.J., KOTHARI, I.L.:** Isolated sieve tubes in the fruit of *Coccinia grandis* (L.) Voight (*Cucurbitaceae*). - Ann. Bot. 51: 251-253, 1983.

12549 - **SHALHEVET, J., HEUER, B., MEIRI, A.:** Irrigation interval as a factor in the salt tolerance of eggplant. - Irrig. Sci. 4: 83-93, 1983.

12550 - **SHALHEVET, J., SHIMSHI, D., MEIR, T.:** Potato irrigation requirements in a hot climate using sprinkler and drip methods. - Agron. J. 75: 13-16, 1983.

*12551 - **SHANMUKHA RAO, S.R., RAMAYYA, N.:** Foliar stomatal abnormalities in the order *Malvales*. - Nat. Acad. Sci. Lett. 5: 283-286, 1982.

12552 - **SHARIFI, M.R.:** The effects of water and nitrogen supply on the competition between three perennial meadow grasses. - Acta oecol. - Oecol. Plant 4: 71-82, 1983.

*12553 - **SHARMA, B.B., PAL, R.K.:** Studies on drip irrigation in grapes. II. Effect on some vegetative growth characters, berry drop and yield. - Ind. J. Agron. 27: 429-434, 1982.

*12554 - **SHARMA, D.P., SHARMA, Y.D., RANA, H.S.:** Stomatal and tree growth characteristics of some crab apples. - Scientia Hort. 17: 327-331, 1982.

*12555 - **SHARMA, H.P., SHOME, U.:** Leaf surface studies of some medicinal salvias. - Proc. Ind. Acad. Sci., Plant Sci. 91: 449-463, 1982.

*12556 - **SHARMA, M.L.:** Soil water regimes and water extraction patterns under two semi-arid shrub (*Atriplex* spp.) communities. - Aust. J. Ecol. 1: 249-258, 1976.

*12557 - **SHARMA, M.L.:** Aspects of salinity and water relations of Australian chenopods. - In: SEN, G.V., RAJPUROHIT, K.S. (ed.): Contributions to the Ecology of Halophytes. Pp. 155-172. Dr W. Junk Publishers, The Hague - Boston - London 1982.

12558 - **SHARMA, R.N., BHARDWAJ, R.B.L.:** Comparative performance of dwarf durum and bread wheats under limited and adequate irrigation supply. - Ind. J. Agron. 28: 37-41, 1983.

12559 - **SHARMA, S.N., RAY, S.B., PANDEY, S.L., PRASAD, R.:** Effect of irrigation, pyrites and phosphobacteria on the efficiency of rock phosphate applied to lentils. - J. agr. Sci. 101: 467-472, 1983.

*12560 - **SHARPE, P.J.H., WU, H.:** Stomatal mechanics: volume changes during opening. - Plant Cell Environ. 1: 259-268, 1978.

12561 - **SHARRATT, B.S., REICOSKY, D.C., IDSO, S.B., BAKER, D.G.:** Relationships between leaf water potential, canopy temperature, and evapotranspiration in irrigated and nonirrigated alfalfa. - Agron. J. 75: 891-894, 1983.

12562 - **SHAW, R.H.:** Estimates of yield reductions in corn caused by water and temperature stress. - In: RAPER, C.D., Jr., KRAMER, P.J. (ed.): Crop Reactions to Water and Temperature Stresses in Humid, Temperate Climates. Pp. 49-65. Westview Press, Boulder 1983.

12563 - **SHCHERBAKOVA, A., KACPERSKA, A.:** Water stress injuries and tolerance as related to potassium efflux from winter rape hypocotyls. - Physiol. Plant. 57: 296-300, 1983.

12564 - **SHEMSANGA, K.H.H., UCHIDA, S., ISHIBASHI. S.:** Influence of moisture content on the mechanical properties of soybeans at cryogenic temperatures. - J. Fac. Agr. Kyushu Univ. 28: 123-134, 1983.

*12565 - **SHEREMET'EV, S.N.:** Intensivnost' transpiratsii i ee vremennaya organizatsiya u muzhskikh i zhenskikh osobeĭ *Pistacia vera* (*Anacardiaceae*). [The transpiration rate and its temporal organization in male and female plants of *Pistacia vera* (*Anacardiaceae*).] - Bot. Zh. 66: 565-572, 1981. [In R.]

*12566 - **SHEVYAKOVA, N.I.:** Soleustoĭchivost' plastomnykh khlorofil'nykh mutantov podsolnechnika. [Salt tolerance in chlorophyll plastid mutants of sunflower.] - Fiziol. Rast. 29: 317-324, 1982. [In R, ab: E.]

12567 - **SHEVYAKOVA, N.I.:** Metabolizm i fiziologicheskaya rol' prolina v rasteniyakh pri vodnom i solevom stresse. [Metabolism and physiological role of proline in plants under conditions of water and salt stress.] - Fiziol. Rast. 30: 768-783, 1983. [In R, ab: E.]

12568 - **SHEW, H.D.:** Effects of soil matric potential on infection of tobacco by *Phytophthora parasitica* var. *nicotianae*. - Phytopathology 73: 1160-1163, 1983.

12569 - **SHIH, S.F., RAHI, G.S.:** Pan evaporation as related to sugarcane leaf area index. - Soil Crop Sci. Soc. Florida Proc. 42: 80-85, 1983.

12570 - **SHIH, S.F., RAHI, G.S., SNYDER, G.H., HARRISON, D.S., SMAJSTRLA, A.G.:** Rice yield, biomass, and leaf area related to evapotranspiration. - Trans ASAE 26: 1458-1464, 1983.

12571 - **SHIMAZAKI, K., GOTOW, K., SAKAKI, T., KONDO, N.:** High respiratory activity of guard cell protoplasts from *Vicia faba* L. - Plant Cell Physiol. 24: 1049 - 1056, 1983.

12572 - **SHIMSHI, D., SHALHEVET, J., MEIR, T.:** Irrigation regime effects on some physiological responses of potato. - Agron. J. 75: 262-267, 1983.

*12573 - **SHINDE, G.G., PAWAR, K.R.:** Effect of irrigation on the pod yield of groundnut. - Ind. J. agr. Sci. 52: 576-578, 1982.

*12574 - **SHMAT'KO, I.G., GULYAEV, B.I., GOLIK, K.N., SHVEDOVA, O.E., LATASHENKO, O.P.:** Vliyanie vodoobespechennosti na gazoobmen kolos'ev i produktivnost' ozimoĭ pshenitsy. [Effect of water supply on gas exchange in ears and productivity of winter wheat.] - Fiziol. Biokhim. kul't. Rast. 14: 286-290, 1983. [In R, ab: E.]

12575 - **SHONE, M.G.T., FLOOD, A.V.:** Effects of periods of localized water stress on subsequent nutrient uptake by barley roots and their adaptation by osmotic adjustment. - New Phytol 94: 561-572, 1983.

12576 - **SHONE, M.G.T., WHIPPS, J.M., FLOOD, A.V.:** Effects of localized and overall water stress on assimilate partitioning in barley between shoots, roots and root exudates. - New Phytol. 95: 625-634, 1983.

*12577 - **SHOUSE, P., JURY, W.A., STOLZY, L.H., DASBERG, S.:** Field measurement and modeling of cowpea water use and yield under stressed and well-watered growth conditions. - Hilgardia 50 (6): 1-25, 1982.

*12578 - **SHUKLA, N.P., HUKKERI, S.B., MENHI LAL:** Effect of irrigations scheduled at different stages of growth on seed production of oat. - Ind. J. agr. Sci. 52: 837-840, 1982.

*12579 - **SIKES, D.A., PETTIT, R.D.:** Soil temperature, oxygen, and water level effects on sand shinnery oak. - Soil Sci. 130: 344-349, 1980.

12580 - **SIKURAJAPATHY, M., CAPPY, J.J., GROSS, H.D.:** A method for inducing controlled moisture stress on seedlings. - Agron. J. 75: 840-843, 1983.

12581 - **SILLS, M.J., CARROW, R.N.:** Turfgrass growth, N use, and water use under soil compaction and N fertilization. - Agron. J. 75: 488-492, 1983.

*12582 - **SILVA, S., VICENTE-CHANDLER, J., ABRUÑA, F.:** Water management systems and rice yields under humid tropical conditions in Puerto Rico. - J. agr. Univ. Puerto Rico 66: 386-389, 1982.

12583 - **SIMINOVITCH, D., CLOUTIER, Y.:** Drought and freezing tolerance and adaptation in plants: some evidence of near equivalences. - Cryobiology 20: 487-503, 1983.

12584 - **ŠIMON, J.:** Působení závlahy, hnojení a výsevku na výnos biomasy ozimé a jarní pšenice. [The effect of irrigation, fertilization and sowing rate on the biomass yield of winter and spring wheat.] - Rost. Výroba (Praha) 29: 827-837, 1983. [In Czech, ab: R, E.]

12585 - **SIMON, P., MARCELLE, R.:** Net photosynthesis and water status of apple seedlings under developing drought conditions: effects of fungicide pre-treatment. - In: MARCELLE, R., CLIJSTERS, H., VAN POUCKE, M. (ed.): Effects of Stress on Photosynthesis. Pp. 133-139. Martinus Nijhoff / Dr W. Junk Publishers, The Hague - Boston - London 1983.

12586 - **SINCLAIR, R., VENABLES, W.N.:** An alternative method for analysing pressure--volume curves produced with the pressure chamber. - Plant Cell Environ. 6: 211-217, 1983.

12587 - **SINCLAIR, T.R., ALLEN, L.H., Jr., COHEN, M.:** Citrus blight effects on carbon dioxide assimilation. - J. Amer. Soc. hort. Sci. 108: 503-506, 1983.

*12588 - **SINGH, A.K., SINHA, A.K., NIMGADE, N.M.:** Effect of evapotranspiration reducing techniques on soil water and temperature status in gram. - Trans. Ind. Soc. Desert Technol. Univ. Cent. Desert Studies 7: 81-83, 1982.

12589 - **SINGH, B.B., GUPTA, D.P.:** Proline accumulation and relative water content in soya bean (*Glycine max*) varieties under water stress. - Ann. Bot. 52: 109-110, 1983.

12590 - **SINGH, B.P.:** Response of mustard and chickpea to moisture in soil profile and plant population on aridisols. - Ind. J. agr. Sci. 53: 543-549, 1983.

*12591 - **SINGH, H.:** Effect of humidity and light periods on infection and sporulation of *Peronospora viciae* on *Pisum sativum*. - Can. J. Bot. 59: 2515-2518, 1981.

12592 - **SINGH, H., ORTON, T.J., HAERTRIDGE-ESH, K.:** Standardization of osmo-conditioning of okra seed. - Ind. J. agr. Sci. 53: 24-26, 1983.

12593 - **SINGH, K.P., MALIK, R.S., MALIK, D.S.:** Diurnal variations in leaf water potential and stomatal conductance of pigeon pea (*Cajanus cajan* (L.) Millsp.) cultivars as affected by irrigation levels. - Biol. Plant. 25: 1-4, 1983.

12594 - **SINGH, P., KANEMASU, E.T.:** Leaf and canopy temperatures of pearl millet genotypes under irrigated and nonirrigated conditions. - Agron. J. 75: 497-501, 1983.

12595 - **SINGH, P., KANEMASU, E.T., SINGH, P.:** Yield and water relations of pearl millet genotypes under irrigated and nonirrigated conditions. - Agron. J. 75: 886-890, 1983.

12596 - **SINGH, S., SINGH, N.P., SINGH, M.:** Influence of irrigation and phosphorus on growth and yield of lentil. - Ind. J. agr. Sci. 53: 225-229, 1983.

12597 - **SINGH, T., MALIK, D.S.:** Effect of water stress at three growth stages on the yield and water-use efficiency of dwarf wheat. - Irrig. Sci. 4: 239-245, 1983.

12598 - **SINGLETON, P.W.:** A split-root growth system for evaluating the effect of salinity on components of the soybean *Rhizobium japonicum* symbiosis. - Crop Sci. 23: 259-262, 1983.

12599 - **SINGLETON, P.W., BOHLOOL, B.B.:** Effect of salinity on the functional components of the soybean-*Rhizobium japonicum* symbiosis. - Crop Sci. 23: 815-818, 1983.

12600 - **SINHA, A.K., PANDEY, S.L.:** Effect of depth to water table on the performance of rice crop. - Ind. J. Agron. 28: 65-69, 1983.

12601 - **SISSON, W.B.:** Carbon balance of *Yucca elata* Engelm. during a hot and cool period *in situ*. - Oecologia 57: 352-360, 1983.

12602 - **SJÖBERG, K., DANELL, K.:** Effects of permanent flooding on *Carex-Equisetum* wetlands in northern Sweden. - Aquat. Bot. 15: 275-286, 1983.

12603 - **SKALABAN, V.D.:** Otsenka produktivnosti sel'skokhozyaĭstvennykh kul'tur po proizvodstvennym faktoram. [Estimation of productivity of agricultural crops from production factors.] - Sel'skokhoz. Biol. 1983 (2): 133-140, 1983. [In R, ab: E.]

12604 - **SKRE, O., OECHEL, W.C., MILLER, P.M.:** Moss leaf water content and solar radiation at the moss surface in a mature black spruce forest in central Alaska. - Can. J. Forest Res. 13: 860-868, 1983.

12605 - **SKRE, O., OECHEL, W.C., MILLER, P.M.:** Patterns of translocation of carbon in four common moss species in a black spruce (*Picea mariana*) dominated forest in interior Alaska. - Can. J. Forest Res. 13: 869-878, 1983.

12606 - **SKRETKOWICZ, A.L., THURTELL, G.W.:** Comparative water stress studies on drought resistant and susceptible corn grown in chamber and field environments. - Can. J. Plant Sci. 63: 775-787, 1983.

*12607 - **SLATER, P.N., JACKSON, R.D.:** Atmospheric effects on radiation reflected from soil and vegetation as measured by orbital sensors using various scanning directions. - Appl. Optics 21: 3923-3931, 1982.

12608 - **SMART, R.E., COOMBE, B.G.:** Water relations of grapevines. - In: KOZLOWSKI, T.T. (ed.): Water Deficits and Plant Growth. Vol. VII. Additional Woody Crop Plants. Pp. 137-196. Academic Press, New York - London - Paris - San Diego - San Francisco - São Paulo - Sydney - Tokyo - Toronto 1983.

*12609 - **SMILLIE, R.M., NOTT, R.:** Salt tolerance in crop plants monitored by chlorophyll fluorescence *in vivo*. - Plant Physiol. 70: 1049-1054, 1982.

12610 - **SMIRNOFF, N., CRAWFORD, R.M.M.:** Variation in the structure and response to flooding of root aerenchyma in some wetland plants. - Ann. Bot. 51: 237-244, 1983.

12611 - **SMITH, J.R.:** The tonoplast impedance of *Chara*. - J. exp. Bot. 34: 120-129, 1983.

12612 - **SMIT-SPINKS, B., SWANSON, B.T., MARKHART, A.H., III.:** Effects of low temperature and light on foliar injury to *Pinus sylvestris* L. - J. Amer. Soc. hort. Sci. 108: 491-495, 1983.

12613 - **SNEED, R.E., PATTERSON, R.P.:** The future role of irrigation in a humid climate. - In: RAPER, C.D., Jr., KRAMER, P.J. (ed.): Crop Reactions to Water and Temperature Stresses in Humid, Temperate Climates. Pp. 187-199. Westview Press, Boulder 1983.

*12614 - **SNELGAR, W.P., BROWN, D.H., GREEN, T.G.A.:** A provisional survey of the interaction between net photosynthetic rate, respiratory rate, and thallus water content in some New Zealand cryptogams. - New Zeal. J. Bot. 18: 247-256, 1980.

*12615 - **SNYDER, R.L.:** Scheduling alfalfa irrigations by computer. - In: Twelfth California Alfalfa Symposium. Pp. 50-52. University of California, Davis 1982.

12616 - **SNYDER, R.L., CARLSON, R.E., SHAW, R.H.:** Leaf-water potential-yield relationships in soybeans. - Iowa State J. Res. 57: 221-232, 1983.

12617 - **SNYDER, R.L., CARLSON, R.E., SHAW, R.H.:** Response of soybean to foliar fertilization and drought stress. - Iowa State J. Res. 57: 259-267, 1983.

12618 - **SOARES, J.M., FARIA, C.M.B., de:** Métodos de irrigação e adubação na cultura do tomate industrial. [Irrigation and fertilization on industrial tomato cultivation.] - Pesq. agropec. Bras. 18: 281-286, 1983. [In Port, ab: E.]

12619 - **SOBRADO, M.A., TURNER, N.C.:** A comparison of the water relations characteristics of *Helianthus annuus* and *Helianthus petiolaris* when subjected to water deficits. - Oecologia 58: 309-313, 1983.

12620 - **SOBRADO, M.A., TURNER, N.C.:** Influence of water deficits on the water relations characteristics and productivity of wild and cultivated sunflowers. - Aust. J. Plant Physiol. 10: 195-203, 1983.

12621 - **SOJKA, R.E., PARSONS, J.E.:** Soybean water status and canopy microclimate relationships at four row spacings. - Agron. J. 75: 961-968, 1983.

12622 - **SOLÁROVÁ, J., POSPÍŠILOVÁ, J.:** Ontogenetické změny závislosti epidermální vodivosti na vodním potenciálu. [Ontogenetic changes in dependence of epidermal conductance on water potential.] - In: Dny Rostlinné Fyziologie III. Pp. 290-291. ČSBS, Praha 1983. [In Czech.]

12623 - **SOLÁROVÁ, J., POSPÍŠILOVÁ, J.:** Photosynthetic characteristics during ontogenesis of leaves. 8. Stomatal diffusive conductance and stomata reactivity. - Photosynthetica 17: 101-151, 1983.

*12624 - **SOLTANI, A., BRIENS, M., GOAS, M.:** Sur le métabolisme glucidique d'*Hedysarum coronarium* L. cultivé en présence de chlorure de sodium. - C. R. Acad. Sci. (Paris) Sér. III 293: 297-300, 1981.

12625 - **SOMERS, D.A., ULLRICH, S.E., RAMSAY, M.F.:** Sunflower germination under simulated drought stress. - Agron. J. 75: 570-572, 1983.

*12626 - **SOMERVILLE, C.R., PORTIS, A.R., Jr., OGREN, W.L.:** A mutant of *Arabidopsis thaliana* which lacks activation of RuBP carboxylase *in vivo*. - Plant Physiol. 70: 381-387, 1982.

12627 - **SOUZA, J.G., de, BARROS, A.C.Q.T., SILVA, J.V., da:** Reservas de hidratos de carbono e resistência do algodoeiro à seca. [Carbohydrate content and drought resistance in cotton.] - Pesq. agropec. Brasil. 18: 269-273, 1983. [In Port, ab: E.]

12628 - **SOUZA, J.G., de, SILVA, J.B.V., da, NETO, M.B., GILES, J.A.:** Velocidade de crescimento da raiz como parâmetro de resistência à seca no algodoeiro. [Speed of growth as a criterion of drought resistance in cotton.] - Pesq. agropec. Bras. 18: 169-172, 1983. [In Port, ab: E.]

*12629 - **SOVONICK-DUNFORD, S., FERRIER, J.M., DAINTY, J.:** Water-relations parameters of the phloem. Determinations of volumetric elastic modulus and membrane conductivity using an applied force method and shrinkage and swelling of tissues in solutions at differing osmotic pressure. - Ann. Bot. 51: 27-37, 1982.

12630 - **SPENCE, R.D., SHARPE, P.J.H., POWELL, R.D., ROGERS, C.A.:** Epidermal and guard cell interactions on stomatal aperture in epidermal strips and intact leaves. - Ann. Bot. 52: 1-12, 1983.

12631 - **SPIERS, J.M.:** Irrigation and peatmoss for the establishment of rabbiteye blueberries. - HortScience 18: 936-937, 1983.

*12632 - **SPITZNER, C., STARK, N.:** Productivity of western larch and subsoil percolation rates on poor andic soils. - Soil Sci. 134: 395-400, 1982.

12633 - **SQUIRE, G.R., BLACK, C.R., ONG, C.K.:** Response to saturation deficit of leaf extension in a stand of pearl millet (*Pennisetum typhoides* S. & H.). II. Dependence on leaf water status and irradiance. - J. exp. Bot. 34: 856-865, 1983.

*12634 - **SQUIRE, G.R., CALLANDER, B.A.:** Tea plantations. - In: KOZLOWSKI, T.T. (ed.): Water Deficits and Plants Growth. Vol. VI. Woody Plant Communities. Pp. 471-510. Academic Press, New York - London - Toronto - Sydney - San Francisco 1981.

12635 - **SQUIRE, G.R., ONG, C.K.:** Response to saturation deficit of leaf extension in a stand of pearl millet (*Pennisetum typhoides* S. & H.). I. Interaction with temperature. - J. exp. Bot. 34: 846-855, 1983.

12636 - **STACK, R.W.:** Take-all of irrigated spring wheat in North Dakota. - North Dakota Farm Res. 40 (6): 25-26, 1983.

12637 - **STANLEY, C.D., HARBAUGH, B.K., PRICE, J.F.:** Environmental factors influencing leaf water potential of *Chrysanthemum*. - J. Amer. Soc. hort. Sci. 108: 237-240, 1983.

12638 - **STAPLETON, A.R.A., WAGENET, R.J., TURNER, D.L.:** Corn growth and nitrogen uptake under irrigated, fertilized conditions. - Irrig. Sci. 4: 1-15, 1983.

12639 - **STEGMAN, E.C.:** Corn and sunflower yield vs management of leaf xylem pressures. - Trans. ASAE 26: 1362-1368, 1374, 1983.

12640 - **STEGMAN, E.C.:** Irrigation options to avoid critical stress: irrigation scheduling - some applied concepts. - In: TAYLOR, H.M., JORDAN, W.R., SINCLAIR, T.R. (ed.): Limitations to Efficient Water Use in Crop Production. Pp. 499-505. Amer. Soc. Agron., Inc., Crop Sci. Soc. Amer., Inc., Soil Sci. Soc. Amer., Inc., Madison 1983.

12641 - **STEINEGGER, D.H., SHEARMAN, R.C., RIORDAN, T.P., KINBACHER, E.J.:** Mower blade sharpness effects on turf. - Agron. J. 75: 479-488, 1983.

12642 - **STEINER, J.L., KANEMASU, E.T., HASZA, D.:** Microclimatic and crop responses to center pivot sprinkler and to surface irrigation. - Irrig. Sci. 4: 201-214, 1983.

12643 - **STEINGRÖVER, E.:** Storage of osmotically active compounds in the taproot of *Daucus carota* L. - J. exp. Bot. 34: 425-433, 1983.

12644 - **STEPONKUS, P.L., DOWGERT, M.F., GORDON-KAMM, W.J.:** Destabilization of the plasma membrane of isolated plant protoplasts during a freeze-thaw cycle: the influence of cold acclimation. - Cryobiology 20: 448-465, 1983.

12645 - **STERN, J., BRESLER, E.:** Nonuniform sprinkler irrigation and crop yield. - Irrig. Sci. 4: 17-29, 1983.

*12646 - **STERN, R.D., DENNETT, M.D., DALE, I.C.:** Analysing daily rainfall measurements to give agronomically useful results. I. Direct methods. - Exp. Agr. 18: 223-236, 1982.

*12647 - **STERN, R.D., DENNETT, M.D., DALE, I.C.:** Analysing daily rainfall measurements to give agronomically useful results. II. A modelling approach. - Exp. Agr. 18: 237-253, 1982.

12648 - **STEUDLE, E., JESCHKE, W.D.:** Water transport in barley roots. Measurements of root pressure and hydraulic conductivity of roots in parallel with turgor and hydraulic conductivity of root cells. - Planta 158: 237-248, 1983.

12649 - **STEUDLE, E., TYERMAN, S.D.:** Determination of permeability coefficients, reflection coefficients, and hydraulic conductivity of *Chara corallina* using the pressure probe: effects of solute concentrations. - J. Membrane Biol. 75: 85-96, 1983.

12650 - **STEUDLE, E., TYERMAN, S.D., WENDLER, S.:** Water relations of plant cells. - In: MARCELLE, R., CLIJSTERS, H., VAN POUCKE, M. (ed.): Effects of Stress on Photosynthesis. Pp. 95-109. Martinus Nijhoff / Dr. W. Junk Publishers, The Hague - Boston - London 1983.

12651 - **STEUDLE, E., ZIEGLER, H., ZIMMERMANN, U.:** Water relations of the epidermal bladder cells of *Oxalis carnosa* Molina. - Planta 159: 38-45, 1983.

*12652 - **STEVANOVIĆ, M., SAVIĆ, R.:** Uticaj količine azota i broja biljaka na prinos hibrida kukuruza u navodnjavanju. [Effect of nitrogen level and plant number on the yield of corn hybrids grown in irrigation.] - Arhiv Poljoprivred. Nauke 43: 455-463, 1982. [In Croat, ab: E.]

*12653 - **STEWART, A.C.:** The effects of high concentrations of salts on photosynthetic electron transport in spinach (*Spinacia oleracea)* chloroplasts. - Biochem. J. 204: 705-712, 1982.

12654 - **STEWART, B.A., MUSICK, J.T., DUSEK, D.A.:** Yield and water use efficiency of grain sorghum in a limited irrigation - dryland farming system. - Agron. J. 75: 629-634, 1983.

12655 - **STEWART, D.W., DWYER, L.M.:** Stomatal response to plant water deficits. - J. theor. Biol. 104: 655-666, 1983.

12656 - **STEWART, J.B.:** A discussion of the relationships between the principal forms of the combination equation for estimating crop evaporation. - Agr. Meteorol. 30: 111-127, 1983.

*12657 - **STOEV, K, SLAVTCHEVA, T.:** La photosynthèse nette chez la vigne (*V. vinifera* L.) et les facteurs écologiques. - Connais. Vigne Vin 16: 171-185, 1982.

12658 - **STRANGEWAYS, I.C.:** Interfacing soil moisture gypsum blocks with a modern data-logging system using a simple, low-cost, DC method. - Soil Sci. 136: 322-324, 1983.

12659 - **STRAUSS, M.S.:** Anatomy and morphology of taro, *Colocasia esculenta* (L.) Schott. - In: WANG, J.K., HIGA, S. (ed.): Taro. A Review of *Colocasia esculenta* and its Potentials. Pp. 20-33. University of Hawaii Press, Honolulu 1983.

12660 - **STUPISHINA, E.A., GUSEV, N.A., CHEREZOV, S.N., TUKHVATULLIN, R.G.**: Gidratatsiya khloroplastov v zavisimosti ot nasyshcheniya ikh vlagoĭ. [Hydration of chloroplasts depending on their saturation with moisture.] - Fiziol. Biokhim kul't. Rast. 15: 183-187, 1983. [In R, ab: E.]

*12661 - **STURTON, S.L., BILANSKI, W.K., MENZIES, D.R.**: Drying of cereal grains with the desiccant bentonite. - Can. agr. Eng. 23: 101-103, 1981.

*12662 - **STUTTE, C.A., da SILVA, P.R.F.**: Nitrogen volatilization from rice leaves. I. Effects of genotype and air temperatures. - Crop Sci. 21: 596-600, 1981.

*12663 - **SUCOFF, E.**: Water relations of the aspens. - Agr. Exp. Sta., Univ. Minnesota Tech. Bull. 338: 1-27, 1982.

12664 - **SUNDBLAD, L.-G., SUNDBOM, E., LUNDGREN, P., JOHANSSON, E.**: Changes in delayed light emission of pine seedlings subjected to water stress. - In: Sixth International Congress on Photosynthesis. Abstracts, Volume 1. Pp. 406. Brussels 1983.

12665 - **SUNELL, L.A., ARDITTI, J.**: Physiology and phytochemistry. - In: WANG, J.-K., HIGA, S. (ed.): Taro. A Review of *Colocasia esculenta* and its Potentials. Pp. 34-140. University of Hawaii Press, Honolulu 1983.

*12666 - **SUSZKA, B., TYLKOWSKI, T.**: Storage of acorns of the English oak (*Quercus robur* L.) over 1 - 5 winters. - Arboretum Kórnickie 25: 199-229, 1980.

12667 - **ŠVIHRA, J.**: Účinky teplotného a vodného stresu na fotosyntézu a tvorbu biomasy. [Effect of heat and water stress on photosynthetic rate and production of biomass.] - In: Dny Rostlinné Fyziologie III. Pp. 294-296. ČSBS, Praha 1983. [In Slov.]

12668 - **SWANEY, D.P., MISHOE, J.W., JONES, J.W., BOGGESS, W.G.**: Using crop models for management: Impact of weather characteristics on irrigation decisions in soybeans. - Trans. ASAE 26: 1808-1814, 1983.

12669 - **SWENSEN, J.B., MURRAY, G.A.**: Cold acclimation of field peas in a controlled environment. - Crop Sci. 23: 27-30, 1983.

12670 - **SWIETLIK, D., KORCAK, R.F., FAUST, M.**: Stomatal and nonstomatal inhibition of photosynthesis in water-stressed apple seedlings. - J. Amer. Soc. hort. Sci. 108: 54-58, 1983.

12671 - **SWIETLIK, D., MILLER, S.S.**: The effect of paclobutrazol on growth and response to water stress of apple seedlings. - J. Amer. Soc. hort. Sci. 108: 1076-1080, 1983.

12672 - **SYVERTSEN, J.P., SMITH, M.L., Jr.**: An inexpensive leaf chamber for measuring net gas exchange. - HortScience 18: 700-701, 1983.

12673 - **SYVERTSEN, J.P., SMITH, M.L., Jr.**: Environmental stress and seasonal changes in proline concentration of citrus tree tissues and juice. - J. Amer. Soc. hort. Sci. 108: 861-866, 1983.

12674 - **SYVERTSEN, J.P., ZABLOTOWICZ, R.M., SMITH, M.L., Jr.**: Soil temperature and flooding effects on two species of citrus. I. Plant growth and hydraulic conductivity. - Plant Soil 72: 3-12, 1983.

12675 - **SZÁBO, I.**: A víztelítettségi hiány vizsgálata négy pázsitfűfaj leveleiben. [Water saturation deficit in the leaves of four forage grasses.] - Növénytermeles 32: 521-529, 1983. [In Hung, ab: E.]

*12676 - **TABBADA, R.A., FLORES, M.A.A.:** Effects of water stress on the microbial populations in the rhizosphere of *Phaseolus vulgaris* L. - Natur. appl. Sci. Bull. 34: 89-95, 1982.

*12677 - **TABBADA, R.A., FLORES, M.A.A.:** Influence of soil water stress on vegetative and reproductive growth of *Phaseolus vulgaris* cv. "White Banguio". - Kalikasan (Philipp. J. Biol.) 11: 266-272, 1982.

12678 - **TABBADA, R.A., FLORES, M.A.A.:** Influence of soil moisture stress on chloroplast ultrastructure in *Psophocarpus tetragonolobus* during early vegetative growth. - Kalikasan (Philipp. J. Biol.) 12: 127-134, 1983.

*12679 - **TAHA, M.A., MALIK, M.N.A., CHAUDHRY, F.I., MAKHDUM, M.I.:** Water requirements of cotton (*Gossypium hirsutum*) sown on different dates in a monsoon climate. - Exp. Agr. 18: 209-217, 1982.

,12680 - **TAINTER, F.H., WILLIAMS, T.M., CODY, J.B.:** Drought as a cause of oak decline and death on the South Carolina coast. - Plant Dis. 67: 195-197, 1983.

12681 - **TAKAHAMA, U.:** Effects of deuterium oxide and free radical scavengers on carotenoid photobleaching in spinach chloroplasts. - Plant Cell Physiol. 24: 495-500, 1983.

12682 - **TAKAOKI, T.:** Accumulation of oxygen gas in excised maize root on aging in water, and water uptake by the root. - Plant Cell Physiol. 24: 1175-1182, 1983.

12683 - **TAKEOKA, Y., KONDO, K., KAUFMAN, P.B.:** Leaf surface fine-structures in rice plants cultured under shaded, and non-shaded conditions. - Jap. J. Crop Sci. 52: 534-543, 1983.

12684 - **TAKEOKA, Y., MATSUMURA, O., KAUFMAN, P.B.:** Studies on silicification of epidermal tissues of grasses as investigated by soft X-ray image analysis. I. On the method to detect and calculate frequency of silica bodies in bulliform cells. - Jap. J. Crop Sci. 52: 544-550, 1983.

*12685 - **TAKIO, S., TAKAOKI, T.:** Light-induced quenching of atebrin fluorescence in relation to H^+ uptake and light-scattering change in isolated spinach chloroplasts. - Photobiochem. Photobiophys. 2: 217-225, 1981.

12686 - **TALALUEVA, L.V., KOSICHENKO, N.E.:** Adaptivnye izmeneniya anatomicheskogo stroeniya list'ev razlichnykh vidov roda *Betula* (*Betulaceae*) v usloviyakh sukhoĭ stepi Nizhnego Povolzh'ya. [Adaptive changes in the anatomical structure of different *Betula* (*Betulaceae*) species under the dry steppe conditions in the Low Volga region.] - Bot. Zh. 68: 1374-1379, 1983. [In R.]

*12687 - **TALHA, M., AMBERGER, A., BURKART, N.:** Effect of soil compaction and soil moisture level on plant growth and potassium uptake. - Z. Acker- Pflanzenbau 148: 156-164, 1979.

12688 - **TAN, C.S., FULTON, J.M., NUTTALL, V.W.:** The influence of soil moisture stress and plant populations on the yield of pickling cucumbers. - Scientia Hort. 21: 217-224, 1983.

12689 - **TANG, Z.C., KOZLOWSKI, T.T.:** Further studies in flood tolerance of *Betula papyrifera* seedlings. - Physiol. Plant. 59: 218-222, 1983.

12690 - **TANG, Z.C., KOZLOWSKI, T.T.:** Responses of *Pinus banksiana* and *Pinus resinosa* seedlings to flooding. - Can. J. Forest Res. 13: 633-639, 1983.

12691 - **TANNER, C.B., SINCLAIR, T.R.:** Efficient water use in crop production: research or re-search? - In TAYLOR, H.M., JORDAN, W.R., SINCLAIR, T.R. (ed.): Limitations to Efficient Water Use in Crop Production. Pp. 1-27. Amer. Soc. Agron., Inc., Crop Sci. Soc. Amer., Inc., Soil Sci. Soc. Amer., Inc., Madison 1983.

12692 - **TANNER, J.E.:** Intracellular diffusion of water. - Arch. Biochem. Biophys. 224: 416-428, 1983.

12693 - **TAYLOR, G.E., Jr., McLAUGHLIN, S.B., Jr., SHRINER, D.S., SELVIDGE, W.J.:** The flux od sulfur-containing gases to vegetation. - Atmos. Environ. 17: 789-796, 1983.

12694 - **TAYLOR, G.E., Jr., TINGEY, D.T.:** Sulfur dioxide flux into leaves of *Geranium carolinianum* L. Evidence for a nonstomatal or residual resistance. - Plant Physiol. 72: 237-244, 1983.

12695 - **TAYLOR, H.M.:** Managing root systems for efficient water use: an overview. - In: TAYLOR, H.M., JORDAN, W.R., SINCLAIR, T.R. (ed.): Limitations to Efficient Water Use in Crop Production. Pp. 87-113. Amer. Soc. Agron., Inc., Crop Sci. Soc. Amer., Inc., Soil Sci. Soc. Amer., Inc., Madison 1983.

12696 - **TAYLOR, H.M., JORDAN, W.R., SINCLAIR, T.R. (ed.):** Limitations to Efficient Water Use in Crop Production. Amer. Soc., Agron., Inc., Crop Sci. Soc. Amer., Inc., Soil Sci. Soc. Amer., Inc., Madison 1983.

12697 - **TAYLOR, H.M., WILLATT, S.T.:** Shrinkage of soybean roots. - Agron. J. 75: 818-820, 1983.

*12698 - **TEAR, E.C., HIGGINBOTHAM, K.O., MAYO, J.M.:** Effects of drying soils on survival of young *Picea glauca* seedlings. - Can. J. Forest Res. 12: 1005-1009, 1982.

*12699 - **TEARE, I.D., PEET, M.M. (ed.):** Crop-Water Relations. - John Wiley & Sons, New York - Chichester - Brisbane - Toronto - Singapore 1982.

12700 - **TELFER, A., ALLEN, J.F., BARBER, J., BENNETT, J.:** Thylakoid protein phosphorylation during State 1-State 2 transitions in osmotically shocked pea chloroplasts. - Biochim. biophys. Acta 722: 176-181, 1983.

12701 - **TERAMURA, A.H., TEVINI, M., IWANZIK, W.:** Effects of ultraviolet-B irradiation on plants during mild water stress. I. Effects on diurnal stomatal resistance. - Physiol. Plant. 57: 175-180, 1983.

12702 - **TERJUNG, W.H., JI, H.-Y., HAYES, J.T., O'ROURKE, P.A., TODHUNTER, P.E.:** Crop water requirements for rainfed and irrigated grain corn in China. - Agr. Water Manage. 6: 43-64, 1983.

*12703 - **TERRES, A.L., GALLI, J., RIBEIRO, A.S.:** Avaliação em arroz para tolerância ao frio. [Screening in rice for cold tolerance.] - Pesq. agropec. Bras. 16: 231-240, 1981. [In Port, ab: E.]

12704 - **TESHA, A.J., ECK, P.:** Effect of nitrogen rate and water stress on growth and water relations of young sweet corn plants. - J. Amer. Soc. hort. Sci. 108: 1049-1053, 1983.

12705 - **TESKEY, R.O., HINCKLEY, T.M., GRIER, C.C.:** Effect of interruption of flow path on stomatal conductance of *Abies amabilis*. - J. exp. Bot. 34: 1251-1259, 1983.

*12706 - **TEŞU, V., TOMA, L.-D., CIREAŞA, E.:** Influenţa unor macro şi microelemente aplicate extraradicular asupra capacităţii de absorbţie a energiei luminoase, a respiraţiei şi a regimului de apă la viţa de vie Aligote. [Effect of extra-root application of macro- and microelements on the capacity of absorption of radiant energy, respiration and water relations in vine cv. Aligoté.] - Lucrari Ştiinţ. Inst. agron. "Ion Ionescu de la Brad" (Iaşi), Ser. Hort. 24: 75-78, 1980. [In Roum, ab: E.]

*12707 - **TEŞU, V., TOMA, L.D., MILICĂ, C.:** Cercetări asupra creşterii şi capacităţii de absorbţie e energiei lumimoase de către pigmenţii asimilatori la floarea scarelui în funcţie de salinitate şi ameliorarea sărăturilor. [Growth and capacity of absorption of light energy by assimilation pigments of sunflower as a function depending on salinity and improvement of saliniferous lands.] - Lucrari Ştiinţ. Inst. agron. "Ion Ionescu de la Brad" (Iaşi), Ser. Agron. 24: 49-52, 1980. [In Roum, ab: E.]

12708 - **TETLOW, R.M.:** Pre-harvest desiccation of crops for conservation. 2. Effect of level of formic acid applied and the volume of solution used on the moisture concentration and chemical composition of a perennial ryegrass sward. - Grass Forage Sci. 38: 27-32, 1983.

12709 - **TEVINI, M., IWANZIK, W., TERAMURA, A.H.:** Effects of UV-B radiation on plants during mild water stress. II. Effects on growth, protein and flavonoid content. - Z. Pflanzenphysiol. 110: 459-467, 1983.

12710 - **TEVINI, M., THOMA, U., IWANZIK, W.:** Effects of enhanced UV-B radiation on germination, seedling growth, leaf anatomy and pigments of some crop plants. - Z. Pflanzenphysiol. 109: 435-448, 1983.

*12711 - **THAKUR, P.S., SINGH, G., RAI, V.K.:** Peroxidase isozymes in relation to developing water deficits in two *Zea mays* L. cultivars. - New Phytol. 89: 25-32, 1981.

*12712 - **THAKUR, R.C., SHEKHAR, J.:** Effect of nitrogen management and soil moisture conservation practices on rainfed wheat (*Triticum aestivum* L.) - Plant Soil 66: 433-437, 1982.

*12713 - **THIEN, S.J.:** The soil as a water reservoir. - In: TEARE, I.D., PEET, M.M. (ed.): Crop-Water Relations. Pp. 45-72. John Wiley & Sons, New York - Chichester - Brisbane - Toronto - Singapore 1982.

12714 - **THOMAS, J.F., HARVEY, C.N.:** Leaf anatomy of four species grown under continuous CO_2 enrichment. - Bot. Gaz. 144: 303-309, 1983.

12715 - **THOMPSON, R.G., TYREE, M.T., LoGULLO, M.A., SALLEO, S.:** The water relations of young olive trees in a mediterranean winter: measurements of evaporation from leaves and water conduction in wood. - Ann. Bot. 52: 399-406, 1983.

12716 - **THUMA, B.A., ROWE, R.C., MADDEN, L.V.:** Relationships of soil temperature and moisture to clubroot (*Plasmodiophora brassicae*) severity on radish in organic soil. - Plant Dis. 67: 758-762, 1983.

12717 - **TIBBITTS, T.W., KOBRIGER, J.M.:** Mode of action of air pollutants in injuring horticultural plants. - HortScience 18: 675-680, 1983.

12718 - **TIESZEN. L.L., DETLING, J.K.:** Productivity of grassland and tundra. - In: LANGE, O.L., NOBEL, P.S., OSMOND, C.B., ZIEGLER, H. (ed.): Physiological Plant Ecology IV. Ecosystem Processes: Mineral Cycling, Productivity and Man's Influence. Pp. 173-203. Springer-Verlag, Berlin - Heidelberg - New York 1983.

*12719 - **TIESZEN, L.L., MILLER, P.C., OECHEL, W.C.:** Photosynthesis. - In: BROWN, J., MILLER, P.C., TIESZEN, L.L., BUNNEL, F.L. (ed.): Arctic Ecosystem. Pp. 102-139, 483-544. Dowden, Hutchinson & Ross, Stroudsburg 1980.

*12720 - **TIETEMA, T.:** Ecophysiology of the sand sedge, *Carex arenaria* L. - Thesis, Rijksuniversiteit Utrecht 1981.

12721 - **TING, I.P., STERNBERG, L.O., DENIRO, M.J.:** Variable photosynthetic metabolism in leaves and stems of *Cissus quadrangularis* L. - Plant Physiol. 71: 677-679, 1983.

12722 - **TITUS, J.E., WAGNER, D.J., STEPHENS, M.D.:** Contrasting water relations of photosynthesis for two *Sphagnum* mosses. - Ecology 64: 1109-1115, 1983.

12723 - **TIVOLI, B., CORBIÈRE, R., JOUAN, B.:** Influence de la température et de l'humidité sur le comportement dans le sol de 3 espèces ou variétés de *Fusarium* responsables de la pourriture sèche des tubercules de pomme terre. - Agronomie 3: 1001-1009, 1983.

12724 - **TKÁČ, J.:** Vplyv dusíkatých hnojív na úrodu siatých tráv v závislosti od vodného režimu pôdy. [The effect of nitrogen fertilizers on the yield of sown grasses in relation to the soil water regime.] - Rost. Výroba (Praha) 29: 847-855, 1983. [In Slov, ab: R, E.]

12725 - **TKACHUK, E.S.:** O vodnom rezhime nekotorykh vidov C_3- i C_4- rastenií. [On water regime of certain C_3- and C_4-plant species.] - Fiziol. Biokhim. kul't. Rast. 15: 460-465, 1983. [In R, ab: E.]

*12726 - **TOKUOKA, M.:** [Water absorption of *Chamaecyparis obtusa* cuttings (III).] - J. Jap. Forest Soc. 59: 118-121, 1977. [In Jap, ab: E.]

*12727 - **TOKUOKA, M.:** [Water absorption of *Chamaecyparis obtusa* cuttings (IV).] - J. Jap. Forest Soc. 59: 178-180, 1977. [In Jap, ab: E.]

12728 - **TOKUOKA, M., TAKEOKA, M.:** [Water uptake and rooting of airlayered *Chamaecyparis obtusa* (I) Experiments using cuttings.] - J. Jap. Forest Soc. 65: 315-320, 1983. [In Jap, ab: E.]

12729 - **TOLLNER, E.W., MOLZ, F.J.:** Simulating plant water uptake in moist, lighter textured soils. - Trans. ASAE 26: 87-91, 99, 1983.

12730 - **TOMBESI, A., DeJONG, T.M., RYUGO, K.:** Net CO_2 assimilation characteristics of walnut leaves under field and laboratory conditions. - J. Amer. Soc. hort. Sci. 108: 558-561, 1983.

12731 - **TOMERLIN, J.R., JONES, A.L.:** Effect of temperature and relative humidity on the latent period of *Venturia inaequalis* in apple leaves. - Phytopathology 73: 51-54, 1983.

*12732 - **TORRES, A.M., DIEDENHOFEN, U.:** Sunflower alcohol dehydrogenase genotypes: germination rates and response to flooding. - Environ. exp. Bot. 21: 35-44, 1981.

12733 - **TORSSELL, B.W.R., KORNHER, A.:** Validation of a yield prediction model for temporary grasslands. - Sweed. J. agr. Res. 13: 125-135, 1983.

*12734 - **TRACHTENBERG, S., ZAMSKI, E.:** The apoplastic conduction of water in *Polytrichum juniperinum* Willd. gametophytes. - New Phytol. 83: 49-52, 1979.

12735 - **TRAVIS, R.L., REED, R.:** The solar tracking pattern in a closed alfalfa canopy. - Crop Sci. 23: 664-668, 1983.

12736 - **TRIGUI, A.:** Evolution des propriétés biophysiques et optiques des jeunes feuilles d'Olivier (*Olea europaea* L.) au cours de leur phase de croissance. - C.R. Acad. Sci. (Paris), Ser. III 297: 543-545, 1983.

*12737 - **TRIVEDI, S., TRIPATHI, R.S.:** The effects of soil texture and moisture on reproductive strategies of *Spergula arvensis* L. and *Plantago major* L. - Weed Res. 22: 41-49, 1982.

12738 - **TROCHOULIAS, T., LAHAV, E.:** The effect of temperature on growth and dry-matter production of macadamia. - Scientia Hort. 19: 167-176, 1983.

*12739 - **TROUGHT, M.C.T., DREW, M.C.:** Effects of waterlogging on young wheat plants (*Triticum aestivum* L.) and on soil solutes at different soil temperatures. - Plant Soil 69: 311-326, 1982.

12740 - **TRUKSA, J., ZÁBORSKÝ, J.:** Vplyv medziriadkovej vzdialenosti na úrodu zrna vybraných hybridov nezavlažovanej a zavlažovanej kukurice. [The effect of inter-row spacing on grain yield in selected hybrids of non-irrigated and irrigated maize.] - Rost. Výroba (Praha) 29: 865-874, 1983. [In Slov, ab: R, E.]

12741 - **TSAREVSKIĬ, Yu.D., IVANNIKOV, V.F., MIRONOVA, N.P.:** K otsenke rasteniĭ ozimoĭ pshenitsy na zasukhoustoĭchivost'. [On estimation of winter wheat plants for drought resistance.] - Sel'skokhoz. Biol. 1983 (3): 35-37, 1983. [In R, ab: E.]

12742 - **TSEL'NIKER, Yu.L., MALKINA, I.S.:** Vliyanie vozrasta dereva na strukturu i funktsiyu fotosinteticheskogo apparata duba chereshchatogo. [Effect of tree age on structure and function of photosynthetic apparatus in english oak.] - Fiziol. Rast. 30: 349-354, 1983. [In R, ab: E.]

12743 - **TSENOV, E.I., KABANOV, V.V., STROGONOV, B.P.:** Vliyanie NaCl na soderzhanie i sintez razlichnykh fraktsiĭ nukleinovykh kislot v apeksakh steblya gorokha. [Effect of NaCl on the content and synthesis of nucleic acids in pea shoot apices.] - Fiziol. Rast. 30: 537-543, 1983. [In R, ab: E.]

*12744 - **TSIPORI, Y., SHIMSHI, D.:** The effect of trickler line spacing on yield of tomatoes (*Lycopersicum esculentum* Mill.). - Soil Sci. Soc. Amer. J. 43: 1225-1228, 1979.

12745 - **TSUNO, Y., SUGIMOTO, H., FUJIYAMA, H., FEILY, I.:** Determination of irrigation-time by the observation of stomatal aperture with the aid of infiltration method. - Jap. J. Crop Sci. 52: 34-42, 1983.

12746 - **TSUNO, Y., TANIYAMA, T., SUPRAPTO, H.:** The photosynthesis and estimation of productivity of cassava. - Jap. J. Crop Sci. 52: 484-492, 1983.

12747 - **TUBELIS, A., SOUZA, F.G. de A.:** Efeito da irrigação na produtividade da cultura do trigo. [Irrigation effect on wheat crop productivity.] - Pesq. agropec. Bras. 18: 1359-1367, 1983. [In Port, ab: E.]

12748 - **TUKHTAEV, A.S.:** Vliyanie vysokoĭ temperatury i degidratatsii na téploustoĭchivost' kletok nekotorykh pustynnykh polukustarnikov. [High temperatures and dehydratation - the effect upon cell thermostability of some desert undershrubs.] - Ékologiya 1983 (1): 34-39, 1983. [In R.]

12749 - **TUMAHAIRWE, J.K., GUMBS, F.A.:** Effect of mulches and irrigation on the production of cabbage (*Brassica oleracea*) in the dry season. - Trop. Agr. 60: 122-127, 1983.

*12750 - **TUR, N.S., KOLESNIKOV, G.P., BRUS, A.G.:** Fotosinteticheskaya produktivnost' sortov risa v usloviyakh zasoleniya. [Photosynthetic productivity of rice cultivars under salinization conditions.] - Byull. nauch.-tekh. Inform. vsesoyuz. nauch.-issled. Inst. Risa (Krasnodar) 28: 20-25, 1980. [In R, ab: E.]

*12751 - **TURNER, F.T., McCAULEY, G.N.:** Rice. - In: TEARE, I.D., PEET, M.M. (ed.): Crop-Water Relations. Pp. 307-350. John Wiley & Sons. New York - Chichester - Brisbane - Toronto - Singapore 1982.

12752 - **TURNER, G.W., LERSTEN, N.R.:** Raised stomatal clusters on *Coleus* (*Lamiaceae*) stems. - Amer. J. Bot. 70: 975-977, 1983.

*12753 - **TURNER, N.C., BURCH, G.J.:** The role of water in plants. - In: TEARE, I.D., PEET, M.M. (ed.): Crop-Water Relations. Pp. 73-126. John Wiley & Sons, New York - Chichester - Brisbane - Toronto - Singapore 1982.

12754 - **TURNER, N.C., SOBRADO, M.A.:** Evaluation of a non-destructive method for measuring turgor pressure in *Helianthus*. - J. exp. Bot. 34: 1562-1568, 1983.

12755 - **TYREE, M.T., GRAHAM, M.E.D., COOPER, K.E., BAZOS, L.J.:** The hydraulic architecture of *Thuja occidentalis*. - Can. J. Bot. 61: 2105-2111, 1983.

12756 - **UDEOGALANYA, A.C.C., MUONEKE, C.O.:** Effect of irrigation levels and pruning heights on the yield of okra (*Hibiscus esculentus* L.) in Southeastern Nigeria. - Beitr. trop. Landwirt. Veterinärmed. 21: 437-444, 1983.

12757 - **UDOVENKO, G.V., DAVYDOVA, G.V.:** Intensivnost' fotosinteza i utilizatsiya assimilyatov u rasteniĭ pshenitsy v usloviyakh zasoleniya. [Photosynthesis intensity and assimilate utilization in wheat plants under saline condiconditions.] - Fiziol. Biokhim. kul't. Rast. 15: 240-245, 1983. [In R, ab: E.]

12758 - **UDOVENKO, G.V., TSIBKOVSKAYA, N.S.:** Dinamika izmeneniĭ ul'trastruktury organell v kletkakh list'ev yachmenya pri zasolenii. [Time course of ultrastructural changes in barley leaf cell organelles under salinity.] - Fiziol. Rast. 30: 518-526, 1983. [In R, ab: E.]

12759 - **ÚLEHLOVÁ, B., ÚLEHLA, J.:** Energy flow in grassland ecosystems. - Ekológia (Bratislava) 2: 37-48, 1983.

*12760 - **UNGAR, I.A.:** Germination ecology of halophytes. - In: SEN, G.V., RAJPUROHIT, K.S. (ed.): Contributions to the Ecology of Halophytes. Pp. 143-154. Dr W. Junk Publishers, The Hague - Boston - London 1982.

12761 - **UNGER, P.W.:** Irrigation effect on sunflower growth, development, and water use. - Field Crops Res. 7: 181-194, 1983.

12762 - **UNGER, P.W., STEWART, B.A.:** Soil management for efficient water use: an overview. - In: TAYLOR, H.M., JORDAN, W.R., SINCLAIR, T.R. (ed.): Limitations to Efficient Water Use in Crop Production. Pp. 419-460. Amer. Soc. Agron., Inc., Crop Sci. Soc. Amer., Inc., Soil Sci. Soc. Amer., Inc., Madison 1983.

12763 - **UPADHYAYA, S.K., RAND, R.H., COOKE, J.R.:** A mathematical model of the effects of CO_2 on stomatal dynamics. - J. theor. Biol. 101: 415-440, 1983.

12764 - **UPCHURCH, D.R., RITCHIE, J.T.:** Root observations using a video recording system in mini-rhizotrons. - Agron. J. 75: 1009-1015, 1983.

12765 - **VÁCLAVÍK, J.:** Vliv přímého ozáření jednotlivých povrchů čepele listů kukuřice na fotosynthetický příjem CO_2. [Effect of direct irradiance on both surfaces of leaf blade on photosynthetic CO_2 influx.] - In: Dny Rostlinné Fyziologie III. Pp. 307-308. ČSBS, Praha 1983. [In Czech.]

12766 - **VALCKE, R., VAN POUCKE, M.:** The effect of water stress on greening of primary barley (*Hordeum vulgare* L. cv. Menuet) leaves. - In: MARCELLE, R., CLIJSTERS, H., VAN POUCKE, M. (ed.): Effects of Stress on Photosynthesis. Pp. 141-150. Martinus Nijhoff / Dr W. Junk Publishers, The Hague - Boston - London 1983.

12767 - **Van ALFEN, N.K., McMILLAN, B.D., TURNER, V., HESS, W.M.:** Role of pit membranes in macromolecule-induced wilt of plants. - Plant Physiol. 73: 1020-1023, 1983.

12768 - **VAN BAVEL, C.H.M., HANKS, R.J.:** Water conservation: principles of soil water flow, evaporation, and evapotranspiration. - In: DREGNE, H.E., WILLIS, W.O. (ed.): Dryland Agriculture. Pp. 25-34. Amer. Soc. Agron., Inc., Crop Sci. Soc. Amer., Inc., Soil Sci. Soc. Amer., Inc., Madison 1983.

12769 - **Van SCHILFGAARDE, J., RAWLINS, S.L.:** Water resources management in a growing society. - In: TAYLOR, H.M., JORDAN, W.R., SINCLAIR, T.R. (ed.): Limitations to Efficient Water Use in Crop Production. Pp. 517-530. Amer. Soc. Agron., Inc., Crop Sci. Soc. Amer., Inc., Soil Sci. Soc. Amer., Inc., Madison 1983.

12770 - **Van STEVENINCK, R.F.M., Van STEVENINCK, M.E.:** Abscisic acid and membrane transport. - In: ADDICOTT, F.T. (ed.): Abscisic Acid. Pp. 171-235. Praeger Publishers, New York 1983.

*12771 - **VAN TUYL, J.M.:** Breeding for resistance to yellow disease of hyacinths. II. Influence of flowering time, leaf characters, stomata and chromosome number on the degree of resistance. - Euphytica 31: 621-628, 1982.

12772 - **VAN VOLKENBURGH, E., DAVIES, W.J.:** Inhibition of light-stimulated leaf expansion by abscisic acid. - J. exp. Bot. 34: 835-845, 1983.

12773 - **VAN VOLKENBURGH, E., HUNT, S., DAVIES, W.J.:** A simple instrument for measuring cell-wall extensibility. - Ann. Bot. 51: 669-672, 1983.

12774 - **Van ZEE, G.A., SCHURER, K.:** On-line estimation of the rate of sap flow in plant stems using stationary thermal response data. - J. exp. Bot. 34: 1636 - 1651, 1983.

12775 - **VAPAAVUORI, E., NURMI, A., KORPILAHTI, E.:** Effect of water stress on rate of photosynthesis, transpiration, and chloroplast ultrastrukture in willow leaves. - In: MARCELLE, R., CLIJSTERS, H., VAN POUCKE, M. (ed.): Effects of Stress on Photosynthesis. Pp. 151-152. Martinus Nijhoff / Dr W. Junk Publishers, The Hague - Boston - London 1983.

12776 - **VARTANIAN, N., WERTHEIMER, D.S., COUDERC, H.:** Scanning electron microscopic aspects of short tuberized roots, with special reference to cell rhizodermis evolution under drought and rehydration. - Plant Cell Environ. 6: 39-46, 1983.

12777 - **VASCONCELLOS, C.A., BARBOSA, J.V.A., DOS SANTOS, H.L., DE FRANÇA, G.E.:** Acumulação de massa seca e de nutrientes por duas cultivares de milho com e sem irrigação suplementar. [Dry matter and nutrient accumulation by two maize cultivars with and without supplemental irrigation.] - Pesq. agropec. Bras. 18: 887-901, 1983. [In Port, ab: E.]

12778 - **VAUGHN, K.C., OUTLAW, W.H., Jr.:** Cytochemical and cytofluorometric evidence for guard cell photosystems. - Plant Physiol. 71: 420-424, 1983.

12779 - **VAVRINA, C.S., ASHLEY, R.A.:** Effect of alachlor on PEG 6000 uptake, root osmotic potential, and root leakage. - Weed Sci. 31: 600-603, 1983.

12780 - **VEDIE, R., ABOU-ZEID, N.:** Rôle de l'eau et des blessures dans l'expression du pouvoir pathogène de *Botrytis fabae* sur *Vicia faba*. - Agronomie 3: 92-93, 1983.

*12781 - **VEERANJANEYULU, K., DAS, V.S.R.:** Photoacoustic spectroscopy - leaf absorption spectra. - J. exp. Bot. 33: 515-519, 1982.

12782 - **VENKATESWARLU, B., RAO, A.V., LAHIRI, A.N.:** Effect of water stress on nodulation and nitrogenase activity of guar (*Cyamopsis tetragonoloba* (L.) Taub.). - Proc. Ind. Acad. Sci., Plant Sci. 92: 297-301, 1983.

*12783 - **VERETENNIKOV, A.V.:** Vliyanie drenazha na protsess fotosinteza drevesnykh rasteniĭ. [Effect of drainage on photosynthesis of woody plants.] - In: Regulyatsiya Fiziologicheskikh Protsessov Rasteniĭ. Pp. 41-47. Izdatel'stvo Voronezhskogo Universiteta, Voronezh 1982. [In R.]

12784 - **VERTUCCI, C.W., LEOPOLD, A.C.:** Dynamics of imbibition by soybean embryos. - Plant Physiol. 72: 190-193, 1983.

*12785 - **VETŐ, F.:** Is the difference of vapour pressure the driving force of real osmotic water transport? - Acta Biochim Biophys. Acad. Sci. Hung. 17: 251-261, 1982.

*12786 - **VETŐ, F.:** A possible driving force for the transport of water in roots: the temperature gradient. - In: BROUWER, R., GASPARÍKOVÁ, O., KOLEK, J., LOUGHMAN, B.C. (ed.): Structure and Function of Plant Roots. Pp. 163-164. Martinus Nijhoff / Dr. W. Junk Publishers, The Hague - Boston - London 1981.

12787 - **VETTERLEIN, E., RÜHL, G., HELLING, R.:** Methodik der Saugspannungsmessung in Druckapparaten: Entwässerungszeiten und Diffusivitäten. - Arch. Acker-Pflanzenbau Bodenk. 27: 79-87, 1983.

12788 - **VICHERKOVÁ, M., CHOVANCOVÁ, A.:** Zastupitelnost draslíkových iontů v reakcích pruduchů ionty sodíkovými. [Possibility of substitution of potassium ions by sodium ions in reactions of stomata.] - In: Dny Rostlinné Fyziologie III. Pp. 311-312. CSBS, Praha 1983. [In Czech.]

*12789 - **VICHEV, Zh., SALDZHIEV, I.:** Fiziko-mekhanichni svoĭstva na khrasta pri novite sortove pamuk v"v vr"zka s mekhaniziranata beritba. [Physico-mechanical characteristics of the cotton bush in new cultivars for mechanical harvesting.] - Rasteniev. Nauki 19 (8): 49-54, 1982. [In Bulg, ab: R, E.]

12790 - **VIDAL, J., GODBILLON, G., GADAL, P.:** Influence of light on phosphoenol pyruvate carboxylase in sorghum leaves. 2. Immunochemical study. - Physiol. Plant. 57: 124-128, 1983.

12791 - **VIEIRA da SILVA, J.:** Environmental conditions and crop characteristics in relation to stress tolerance. - In: RAPER, C.D., Jr., KRAMER, P.J. (ed.): Crop Reactions to Water and Temperature Stresses in Humid, Temperate Climates. Pp. 341-346. Westview Press, Boulder 1983.

12792 - **VIEIRA da SILVA, J., ROUAULT, O., KPAVODE, H., PHAM THI, A.:** Effect of drought on ribulose bisphosphate carboxylase/oxygenase activity in soya bean and oil palm leaf tissues. - In: Sixth International Congress on Photosynthesis. Abstracts, Volume 1. Pp. 408. Brussels 1983.

12793 - **VINCENT, J.F.V.:** The influence of water content on the stiffness and fracture properties of grass leaves. - Grass Forage Sci. 38: 107-114, 1983.

*12794 - **VIRZO DE SANTO, A., ALFANI, A., FIORETTO, A.:** Relazioni tra CAM e grado di xeromorfismo in alcune *Peperomie* succulente. [Relationships between CAM and the degree of xeromorphism in some succulent species of *Peperomia.*] - Delpinoa 20: 15-29, 1978. [In Ital, ab: E.]

12795 - **VIRZO DE SANTO, A., ALFANI, A., RUSSO, G., FIORETTO, A.:** Relationship between CAM and succulence in some species of *Vitaceae* and *Piperaceae*. - Bot. Gaz. 144: 342-346, 1983.

*12796 - **VIRZO DE SANTO, A., FIORETTO, A., ALFANI, A.:** Resistenza alla siccità di *Cissus quadrangularis* L., Vitacea succulenta con CAM. [Drought resistance of *Cissus quadrangularis* L., a succulent of the *Vitaceae* with CAM.] - Delpinoa 20: 37-44, 1978. [In Ital, ab: E.]

*12797 - **VLAKHOVA, M.:** Prouchvane samostoyatelnoto i s"vmestnoto vliyanie na napoyavaneto i g"stotata na poseva v"rkhu razvitieto i dobiva na ranni khibridi tsarevitsa za z"rno v raĭona na yugoiztochna B"lgariya. [The single and combined effect of irrigation and plant density on the development and yield of early maize hybrids in the region of south-eastern Bulgaria.] - Rasteniev. Nauki 19 (8): 17-25, 1982. [In Bulg, ab: R, E.]

*12798 - **VOLKOVA, N.P., SMETANIN, A.P.:** Reaktsiya perspektivnykh sortov risa na glubokiĭ sloĭ vody v nachal'nye fazy vegetatsii. [Reaction of promising rice varieties to deep water layer during the initial stages of development.] - Byull. nauch.-tekh. Informatsii vsesoyuz. nauch.-issled. Inst. Risa (Krasnodar) 27: 3-6, 1980. [In R, ab: E.]

*12799 - **VOROB'EV, N.V., ALESHIN, E.P.:** Vskhozhest' semyan i dlina koleoptileĭ u prorostkov pri prorastanii risa v zatoplennoĭ srede pri raznoĭ temperature. [Seed germination and the coleoptiles length of the seedlings of rice germinating in the flooded substrate under various temperatures.] - Byull. nauch.-tekh. Informatsii vsesoyuz. nauch.-issled. Inst. Risa 28: 26-30, 1980. [In R, ab: E.]

12800 - **WAGENET, R.J., RODRIGUEZ, R.R., CAMPBELL, W.F., TURNER, D.L.:** Fertilizer and salty water effects on *Phaseolus*. - Agron. J. 75: 161-166, 1983.

*12801 - **WAGHMODE, A.P., JOSHI, G.V.:** Chemical composition of leaves of halophytes & sediments in estuarine habitat. - Ind. J. mar. Sci. 11: 104-106, 1982.

*12802 - **WALDREN, R.P.:** Corn. - In: TEARE, I.D., PEET, M.M. (ed.): Crop-Water Relations. Pp. 187-212. John Wiley & Sons, New York - Chichester - Brisbane - - Toronto - Singapore 1982.

12803 - **WALKER, G.K., HATFIELD, J.L.:** Stress measurement using foliage temperatures. - Agron. J. 75: 623-629, 1983.

12804 - **WALKER, R.R., DOUGLAS, T.J.:** Effect of salinity level on uptake and distribution of chloride, sodium and potassium ions in citrus plants. - Aust. J. agr. Res. 34: 145-153, 1983.

12805 - **WALKER, R.R., TÖRÖKFALVY, E., GRIEVE, A.M., PRIOR, L.D.:** Water relations and ion concentrations of leaves on salt-stressed citrus plants. - Aust. J. Plant Physiol. 10: 265-277, 1983.

12806 - **WALKER, S., OOSTERHUIS, D.M., SAVAGE, M.J.:** Field use of screen-caged thermocouple psychrometers in sample chambers. - Crop Sci. 23: 627-632, 1983.

12807 - **WALLACE, J.S., BISCOE, P.V.:** Water relations of winter wheat. 4. Hydraulic resistance and capacitance in the soil-plant system. - J. agr. Sci. 100: 591-600, 1983.

12808 - **WALLACE, J.S., CLARK, J.A., McGOWAN, M.:** Water relations of winter wheat. 3. Components of leaf water potential and the soil-plant water potential gradient. - J. agr. Sci. 100: 581-589, 1983.

*12809 - **WALLNER, S.J., BECWAR, M.R., BUTLER, J.D.:** Measurement of turfgrass heat tolerance *in vitro*. - J. Amer. Soc. hort. Sci. 107: 608-613, 1982.

12810 - **WALTER, H., BRECKLE, S.-W.:** Ökologische Grundlagen in Globaler Sicht. - Gustav Fischer-Verlag, Stuttgart 1983.

12811 - **WAMPLE, R.L., DAVIS, R.W.:** Effect of flooding on starch accumulation in chloroplasts of sunflower (*Helianthus annuus* L.). - Plant Physiol. 73: 195-198, 1983.

12812 - **WAMPLE, R.L., FARRAR, S.L.:** Yield and quality of furrow and trickle irrigated hop (*Humulus lupulus* L.) in Washington State. - Agr. Water Manage. 7: 457-470, 1983.

12813 - **WANG, J.-K., HIGA, S. (ed.):** Taro. A Review of *Colocasia esculenta* and its Potentials. - University of Hawaii Press, Honolulu 1983.

12814 - **WARD, F.H., POWELL, A.A.:** Evidence for repair processes in onion seeds during storage at high seed moisture contents. - J. exp. Bot. 34: 277-282, 1983.

12815 - **WARDLE, K., DOBBS, E.B., SHORT, K.C.:** *In vitro* acclimatization of aseptically cultured plantlets to humidity. - J. Amer. Soc. hort. Sci. 108: 386-389, 1983.

12816 - **WARDLE, K., SHORT, K.C.:** Stomatal responses and the senescence of leaves. - Ann. Bot. 52: 411-412, 1983.

12817 - **WARDLE, K., SHORT, K.C.:** Stomatal response of *in vitro* cultured plantlets. I. Responses in epidermal strips of *Chrysanthemum* to environmental factors and growth regulators. - Biochem. Physiol. Pflanz. 178: 619-624, 1983.

12818 - **WARRICK, A.W.:** Interrelationships of irrigation uniformity terms. - J. Irrig. Drain Eng. 109: 317-332, 1983.

12819 - **WARRICK, A.W., GARDNER, W.R.:** Crop yield as affected by spatial variations of soil and irrigation. - Water Resour. Res. 19: 181-186, 1983.

12820 - **WATT, T.A.:** The effects of salt water and soil type upon the germination, establishment and vegetative growth of *Holcus lanatus* L. and *Lolium perenne* L. - New Phytol. 94: 275-291, 1983.

12821 - **WEAVER, S.E., TAN, C.S.:** Critical period of weed interference in transplanted tomatoes (*Lycopersicon esculentum*): Growth analysis. - Weed Sci. 31: 476-481, 1983.

*12822 - **WEBB, M.A., ARNOTT, H.J.:** Cell wall conformation in dry seeds in relation to the preservation of structural integrity during desiccation. - Amer. J. Bot. 69: 1657-1668, 1982.

12823 - **WEBB, W.L., LAUENROTH, W.K., SZAREK, S.R., KINERSON, R.S.:** Primary production and abiotic controls in forests, grasslands and desert ecosystems in the United States. - Ecology 64: 134-151, 1983.

12824 - **WEI, J.-M., SHEN, Y.-G., LI, D.-Y.:** [Studies on the coupling mechanism of photophosphorylation. VI. Effect of the integrity of isolated chloroplasts on P/e_2 ratio.] - Acta Phytophys. Sin. 9: 231-240, 1983. [In Chin, ab: E.]

*12825 - **WEINBAUM, S.A., EVEN-CHEN, Z., KESTER, D.E.:** Increased stomatal resistance in two cultivars of almond sensitive to bud failure. - HortScience 15: 583-585, 1980.

12826 - **WEISS, A.:** A quantitative approach to the Pruitt and Doorenbos version of the Penman equation. - Irrig. Sci. 4: 267-275, 1983.

12827 - **WEISS, A., HAGEN, A.F.:** Further experiments on the measurement of leaf wetness. - Agr. Meteorol. 29: 207-212, 1983.

12828 - **WENKERT, W.:** Water transport and balance within the plant: an overview. - In: TAYLOR, H.M., JORDAN, W.R., SINCLAIR, T.R. (ed.): Limitations to Efficient Water Use in Crop Production. Pp. 137-172. Amer. Soc. Agron., Inc., Crop Sci. Soc. Amer., Inc., Soil Sci. Soc. Amer., Inc., Madison 1983.

12829 - **WERK, K.S., EHLERINGER, J., FORSETH, I.N., COOK, C.S.:** Photosynthetic characteristics of Sonoran Desert winter annuals. - Oecologia 59: 101-105, 1983.

12830 - **WERKER, E., LERNER, H.R., WEIMBERG, R., POLJAKOFF-MAYBER, A.:** Structural changes occurring in nuclei of barley root cells in response to a combined effect of salinity and ageing. - Amer. J. Bot. 70: 222-225, 1983.

*12831 - **WERTHEIMER, D.S., VARTANIAN, N.:** Sur l'absence de poils absorbants des racines courtes et tubérisées formées en conditions de sécheresse chez le *Sinapis alba*. - Ann. Sci. Natur. Bot. 13e Sér. 2: 1-9, 1981.

12832 - **WETZSTEIN, H.Y., SOMMER, H.E.:** Scanning electron microscopy of *in vitro*-cultured *Liquidambar styraciflua* plantlets during acclimatization. - J. Amer. Soc. hort. Sci. 108: 475-480, 1983.

12833 - **WEYERS, J.D.B., FITZSIMONS, P.J., MANSEY, G.M., MARTIN, E.S.:** Guard cell protoplasts - Aspects of work with an important new research tool. - Physiol. Plant. 58: 331-339, 1983.

12834 - **WHALE, D.M.:** The response of *Primula* species to soil waterlogging and soil drought. - Oecologia 58: 272-277, 1983.

12835 - **WHALE, D.M.:** Seasonal variation in the gas exchange characteristics of *Primula* species. - Oecologia 59: 377-383, 1983.

12836 - **WHITEHEAD, D., SHERIFF, D.W., GREER, D.H.:** The relationship between stomatal conductance, transpiration rate and tracheid structure in *Pinus radiata* clones grown at different water vapour saturation deficits. - Plant Cell Environ. 6: 703-710, 1983.

*12837 - **WHITELEY, G.M., HEWITT, J.S., DEXTER, A.R.:** The buckling of plant roots. - Physiol. Plant. 54: 333-342, 1982.

12838 - **WIEGERT, R.G., CHALMERS, A.G., RANDERSON, P.F.:** Productivity gradients in salt marshes: the response of *Spartina alterniflora* to experimentally manipulated soil water movement. - Oikos 41: 1-6, 1983.

12839 - **WIEGHART, M., THOMAS, J.W., TESAR, M.B., HANSEN, C.M.:** Acceleration of alfalfa drying in the field by chemical application at cutting. - Crop Sci. 23: 225-229, 1983.

12840 - **WIENCKE, C., LÄUCHLI, A.:** Tonoplast fine structure and osmotic regulation in *Porphyra umbilicalis*. - Planta 159: 342-346, 1983.

12841 - **WIERSUM, L.K., HARMANNY, K.:** Changes in the water-permeability of roots of some trees during drought stress and recovery, as related to problems of growth in urban environment. - Plant Soil 75: 443-448, 1983.

12842 - **WIERTZ, R.:** Verbesserung der Wasserverbrauchsmessung in einer Lysimeteranlage. - Gartenbauwissenschaft 48: 188, 1983.

*12843 - **WIJK, A.L.M. van, FEDDES, R.A.:** A model approach to the evaluation of drainage effects. - Techn. Bull. New. Ser. 15: 131-149, 1982.

12844 - **WILCOX, D.A., DAVIES, F.S., BUCHANAN, D.W.:** Root temperature, water relations, and cold hardiness in two citrus rootstocks. - J. Amer. Soc. hort. Sci. 108: 318-321, 1983.

12845 - **WILLERT, D.J. von, BRINCKMANN, E., BAASCH, R., ELLER, B.M.:** *Welwitschia mirabilis* Hook. Fil. - a CAM plant? Ecological investigations in the central Namib desert. - In: MARCELLE, R., CLIJSTERS, H., VAN POUCKE, M. (ed.): Effects of Stress on Photosynthesis. Pp. 185-191. Martinus Nijhoff / Dr W. Junk Publishers, The Hague - Boston - London 1983.

12846 - **WILLERT, D.J. von, BRINCKMANN, E., ELLER, B.M., SCHEITLER, B.:** CO_2 exchange of CAM exhibiting succulents in the southern Namib desert in relation to microclimate and water stress. - Photosynthesis Res. 4: 289-298, 1983.

12847 - **WILLERT, D.J. von, BRINCKMANN, E., SCHEITLER, B., ELLER, B.M.:** Responses of Crassulacean Acid Metabolism (CAM) to increasing and decreasing water stress in plants in the southern Namib desert. - In: MARCELLE, R., CLIJSTERS, H., VAN POUCKE, M. (ed.): Effects of Stress on Photosynthesis. Pp. 155-163. Martinus Nijhoff / Dr W. Junk Publishers, The Hague - Boston - London 1983.

*12848 - **WILLIAMS, P.C., NORRIS, K.H., ZAROWSKI, W.S.:** Influence of temperature on estimation of protein and moisture in wheat by near-infrared reflectance. - Cereal. Chem. 59: 473-477, 1982.

12849 - **WILLIAMS, W.E.:** Optimal water-use efficiency in a California shrub. - Plant Cell Environ. 6: 145-151, 1983.

12850 - **WILLIAMS, W.E., GRIVET, C., ZEIGER, E.:** Gas exchange in *Paphiopedilum*. Lack of chloroplasts in guard cells correlates with low stomatal conductance. - Plant Physiol. 72: 906-908, 1983.

12851 - **WILLMER, C.M.:** Phosphoenolpyruvate carboxylase activity and stomatal operation. - Physiol. vég. 21: 943-953, 1983.

12852 - **WILLMER, C.M., RUTTER, J.C., MEIDNER, H.:** Potassium involvement in stomatal movements of *Paphiopedilum*. - J. exp. Bot. 34: 507-513, 1983.

12853 - **WILSON, J.M.:** Interaction of chilling and water stress. - In: RAPER, C.D., Jr., KRAMER, P.J. (ed.): Crop Reactions to Water and Temperature Stresses in Humid, Temperate Climates. Pp. 133-147. Westview Press, Boulder 1983.

12854 - **WILSON, J.R.:** Effects of water stress on *in vitro* dry matter digestibility and chemical composition of herbage of tropical pasture species. - Aust. J. agr. Res. 34: 377-390, 1983.

12855 - **WILSON, J.R., BROWN, R.H., WINDHAM, W.R.:** Influence of leaf anatomy on the dry matter digestibility of C_3, C_4, and C_3/C_4 intermediate types of *Panicum* species. - Crop Sci. 23: 141-146, 1983.

12856 - **WILSON, J.R., LUDLOW, M.M.:** Time trends for change in osmotic adjustment and water relations of leaves of *Cenchrus ciliaris* during and after water stress. - Aust. J. Plant Physiol. 10: 15-24, 1983.

12857 - **WILSON, J.R., LUDLOW, M.M.:** Time trends of solute accumulation and the influence of potassium fertilizer on osmotic adjustment of water-stressed leaves of three tropical grasses. - Aust. J. Plant Physiol. 10: 523-537, 1983.

12858 - **WINKLER, F.J., SCHMIDT, H.-L., WIRTH, E., LATZKO, E., LENHART, B., ZIEGLER, H.:** Temperature, pH and enzyme-source dependence of the HCO_3^- carbon isotope effect on the phosphoenolpyruvate carboxylase reaction. - Physiol. vég. 21: 889-895, 1983.

12859 - **WINNER, W.E., BEWLEY, J.D.:** Photosynthesis and respiration of feather mosses fumigated at different hydration levels with SO_2. - Can. J. Bot. 61: 1456-1461, 1983.

*12860 - **WINNER, W.E., KOCH, G.W.:** Water relations and SO_2 resistance of mosses. - J. Hattori bot. Lab. 1982 (52): 431-440, 1982.

12861 - **WINTER, K., DEMMIG, B.:** Chloroplasts from *Mesembryanthemum crystallinum*, a halophilic plant capable of Crassulacean acid metabolism. - In: Sixth International Congress on Photosynthesis. Abstracts, Volume 1. Pp. 410. Brussels 1983.

12862 - **WISE, R.R., McWILLIAM, J.R., NAYLOR, A.W.:** A comparative study of low-temperature-induced ultrastructural alterations of three species with differing chilling sensitivities. - Plant Cell Environ. 6: 525-535, 1983.

12863 - **WITTENBACH, V.A.:** Effect of pod removal on leaf photosynthesis and soluble protein composition of field-grown soybeans. - Plant Physiol. 73: 121-124, 1983.

*12864 - **WITTENBACH, V.A.:** Effect of pod removal on leaf senescence in soybeans. - Plant Physiol. 70: 1544-1548, 1982.

12865 - **WITTWER, S.H.:** Rising atmospheric CO_2 and crop productivity. - HortScience 18: 667-673, 1983.

*12866 - **WOLF, D.D., PARRISH, D.J.:** Short-term growth responses of tall fescue to changes in soil water potential and to defoliation. - Crop Sci. 22: 996-999, 1982.

12867 - **WOLFE, D.W., FERERES, E., VOSS, R.E.:** Growth and yield response of two potato cultivars to various levels of applied water. - Irrig. Sci. 3: 211-222, 1983.

12868 - **WONG, P.T.W.:** Effect of osmotic potential on the growth of *Gaeumannomyces graminis* and *Phialophora* spp. - Ann. appl. Biol. 102: 67-78, 1983.

12869 - **WOODHOUSE, R.M., WILLIAMS, J.G., NOBEL, P.S.:** Simulation of plant temperature and water loss by the desert succulent, *Agave deserti*. - Oecologia 57: 291-297, 1983.

*12870 - **WOODSTOCK, L.W., TAYLORSON, R.B.:** Ethanol and acetaldehyde in imbibing soybean seeds in relation to deterioration. - Plant Physiol. 67: 424-428, 1981.

*12871 - **WOODSTOCK, L.W., TAYLORSON, R.B.:** Soaking injury and its reversal with polyethylene glycol in relation to respiratory metabolism in high and low vigor soybean seeds. - Physiol. Plant. 53: 263-268, 1981.

12872 - **WOOLHOUSE, H.W.:** The effects of stress on photosynthesis. - In: MARCELLE, R., CLIJSTERS, H., VAN POUCKE, M. (ed.): Effects of Stress on Photosynthesis. Pp. 1-28. Martinus Nijhoff / Dr W. Junk Publishers, The Hague - Boston - London 1983.

12873 - **WOOLLEY, J.T.:** Direct microscopic observation of stomata in the field - quick and easy. - Crop Sci. 23: 589-590, 1983.

*12874 - **WORLEY, R.E., LITTRELL, R.H.:** Factors affecting injection rate for pressure trunk injection of pecan trees. - Scientia Hort. 14: 323-328, 1981.

* 12875 - **WORWOOD, D.R., WALKER, D.R., WILLARDSON, L.S.:** The effect of three levels of trickle irrigation on consumptive use and depth water extraction correlated with peach tree growth and yield. - HortScience 17: 580, 1982.

12876 - **WRIGHT, G.C., SMITH, R.C.G.:** Differences between two grain sorghum genotypes in adaptation to drought stress. II. Root water uptake and water use. - Aust. J. agr. Res. 34: 627-636, 1983.

12877 - **WRIGHT, G.C., SMITH, R.C.G., McWILLIAM, J.R.:** Differences between two grain sorghum genotypes in adaptation to drought stress. I. Crop growth and yield responses. - Aust. J. agr. Res. 34: 615-626, 1983.

12878 - **WRIGHT, G.C., SMITH, R.C.G., MORGAN, J.M.:** Differences between two grain sorghum genotypes in adaptation to drought stress. III. Physiological responses. - Aust. J. agr. Res. 34: 637-651, 1983.

12879 - **WRIGHT, J.P., FISHER, D.B.:** Estimation of the volumetric elastic modulus and membrane hydraulic conductivity of willow sieve tubes. - Plant Physiol. 73: 1042-1047, 1983.

*12880 - **WU, H., SHARPE, P.J.H.:** Stomatal mechanics II: material properties of guard cell walls. - Plant Cell Environ. 2: 235-244, 1979.

*12881 - **WU, L.:** The potential for evolution of salinity tolerance in *Agrostis stolonifera* L. and *Agrostis tenuis* Sibth. - New Phytol. 89: 471-486, 1981.

12882 - **WU, L., HUFF, D.R.:** Characteristics of creeping bentgrass clones (*Agrostis stolonifera* L.) from a salinity-tolerant population after surviving drought stress. - HortScience 18: 883-885, 1983.

12883 - **WURST, F., HOCHE, C., BANCHER, E.:** Veränderungen des Phytosterolgehaltes in *Digitalis lanata* unter Wasserstress. - Phyton 23: 91-99, 1983.

12884 - **WUTSCHER, H.K.:** Growth, mineral uptake, and evapotranspiration of air layers from blight-affected and healthy grapefruit trees with two nutrient regimes. - J. Amer. Soc. hort. Sci. 108: 84-87, 1983.

12885 - **YADAV, R.C., KUMAR, V., OGUNLELA, V.B.:** Biological water harvesting: a method of enabling dryland crops to endure periods of drought. - J. Arid Environ. 6: 115-117, 1983.

*12886 - **YAIR, A., SHACHAK, M.:** A case study of energy, water and soil flow chains in an arid ecosystem. - Oecologia 54: 389-397, 1982.

12887 - **YAMAMOTO, T.:** Models of water competition between fruits and leaves on spurs of "Bartlett" pear trees and its measurement by a heat-pulse method. - Scientia Hort. 20: 241-250, 1983.

12888 - **YAMAMOTO, T., WATANABE, S.:** [Water potentials and water fluxes into fruit at the initial time of development of hard end disorder in 'Bartlett' pear.] - J. Jap. Soc. hort. Sci. 51: 395-404, 1983. [In Jap, ab: E.]

12889 - **YČAS, J.W., ZOBEL, R.W.:** The response of maize radicle orientation to soil solution and soil atmosphere. - Plant Soil 70: 27-35, 1983.

12890 - **YEO, A.R.:** Salinity resistance: Physiologies and prices. - Physiol. Plant. 58: 214-222, 1983.

*12891 - **YOUNG, D., SHERALI, H.D.:** Constructive derivation in dynamic programing modeling. - Water Resour. Res. 17: 293-294, 1981.

12892 - **YOUNG, D.R., SMITH, W.K.:** Effect of cloudcover on photosynthesis and transpiration in the subalpine understory species *Arnica latifolia*. - Ecology 64: 681-687, 1983.

*12893 - **ZABUGA, V.F., SHCHERBATYUK, A.S.:** Ėkologiya fotosinteza sosny obyknovennoĭ lesostepnogo Predbaĭkal'ya. [Ecology of photosynthesis of *Pinus silvestris* in the forest-steppe of pre-Baikal region.] - Ėkologiya 1982 (5): 76-78, 1982. [In R.]

12894 - **ZAERR, J.B.:** Short-term flooding and net photosynthesis in seedlings of three conifers. - Forest Sci. 29: 71-78, 1983.

*12895 - **ZAJĄCZKOWSKA, J., ŁOTOCKI, A., MORTECZKA, H., WITKOWSKA, D.:** Ageing of assimilatory organs of conifers under natural and polluted environmental conditions. - Polish ecol. Stud. 7: 401-413, 1981.

12896 - **ZÁMEČNÍK, J.:** Závislost hodnot vodního sytostního deficitu na teplotě během sycení terčíků. [Dependence of water saturation deficit values on temperature during saturation.] - In: Dny Rostlinné Fyziologie III. Pp. 313-314. ČSBS, Praha 1983. [In Czech]

12897 - **ZEEVAART, J.A.D.:** Metabolism of abscisic acid and its regulation in *Xanthium* leaves during and after water stress. - Plant Physiol. 71: 477-481, 1983.

12898 - **ZEIGER, E.:** The biology of stomatal guard cells. - Annu. Rev. Plant Physiol. 34: 441-475, 1983.

12899 - **ZEIGER, E., ASSMANN, S.M., MEIDNER, H.:** The photobiology of *Paphiopedilum* stomata: opening under blue but not red light. - Photochem. Photobiol. 38: 627-630, 1983.

*12900 - **ŻELAWSKI, W., ZAJĄCZKOWSKA, J., MORTECZKA, H.:** Photosynthesis of young and old trees of Scots pine (*Pinus silvestris* L.) in relation to ageing phenomena. - Polish ecol. Stud. 7: 453-462, 1981.

12901 - **ZEMÁNEK, M.:** Reakce genotypů jarního ječmene na rozdílné podmínky zásobení vodou a minerálními živinami. [The response of spring barley genotypes to different water and mineral nutrient regimes.] - Rost. Výroba (Praha) 29: 633-642, 1983. [In Czech, ab: R, E.]

12902 - **ZHAO, Y.J., KAMISAKA, S., MASUDA, Y.:** Osmoregulation in hypocotyls of etiolated mung bean seedlings with or without cotyledons in response to water-deficient stress. - Bot. Mag. (Tokyo) 96: 211-222, 1983.

12903 - **ZIKMANIS, P.B., AUZĀNE, S.I., KRŪĈE, R.V., AUZINA, L.P., BEKER, M.J.:** Interrelationships between the fatty acid composition and metabolic pathways upon dehydration-rehydration of the yeast *Saccharomyces cerevisiae*. - Europ. J. appl. Microbiol. Biotechnol. 18: 298-302, 1983.

12904 - **ZIMA, M.:** Rýchlosť fotosyntézy a transpirácie ako kritérium suchovzdornosti. [Photosynthetic and transpiration rate as indicator for drought resistance.] - In: Dny Rostlinné Fyziologie III. Pp. 317-318. ČSBS, Praha 1983. [In: Slov.]

12905 - **ZIMMERMANN, M.H.:** Xylem Structure and the Ascent of Sap. - Springer-Verlag, Berlin - Heidelberg - New York - Tokyo 1983.

12906 - **ZIMMERMANN, M.H., SPERRY, J.S.:** Anatomy of the palm *Rhapis excelsa*, IX. Xylem structure of the leaf insertion. - J. Arnold Arboretum 64: 599-609, 1983.

12907 - **ZISKA, L.H., HALL, A.E.:** Seed yields and water use of cowpeas (*Vigna unguiculata* (L.) Walp.) subjected to planned-water-deficit irrigation. - Irrig. Sci. 3: 237-245, 1983.

12908 - **ZISKA, L.H., HALL, A.E.:** Soil and plant measurements for determining when to irrigate cowpeas (*Vigna unguiculata* (L.) Walp.) grown under planned-water--deficits. - Irrig. Sci. 3: 247-257, 1983.

12909 - **ZOBEL, R.W.:** Crop manipulation for efficient use of water: constraints and potential techniques in breeding for efficient water use. - In: TAYLOR, H.M., JORDAN, W.R., SINCLAIR, T.R. (ed.): Limitations to Efficient Water Use in Crop Production. Pp. 381-392. Amer. Soc. Agron., Inc., Crop Sci. Soc. Amer., Inc., Soil Sci. Soc. Amer., Inc., Madison 1983.

12910 - **ZOLG, M., BORNKAMM, R.:** Über die Auswirkung von Streusalz auf die Alterung der Blätter verschiedener Strassenbaumarten. - Flora 174: 377-404, 1983.

12911 - **ZUR, B., JONES, J.W., BOOTE, K.J.:** Field evaluation of a water relations model for soybean. I. Validity of some basic assumptions. - Agron. J. 75: 272-280, 1983.

AUTHORS' INDEX

Authors' names are presented in the form in which they appear in the respective publication. The names from papers published in Cyrillic character are transcribed as shown in Instructions for Use. Alternative spelling and form of the name of the same author are usually cross-indexed.

A

B

D

E

G

H

I

J

K

L

M

N

O

Q

U

V

W

X

Y

Z

PLANT INDEX

This index contains plant genera and types interesting as experimental material for physiological, ecological and agricultural studies. The Latin plant names are the main items which present the reference number. English names of the most common plants are cross-indexed.

C

Q

R

SUBJECT INDEX

This index contains a selection of primary items chosen according to their interest for water relation researchers and to their relative importance and occurrence.

A

(continued)

D

E

F

G

H

I

L

M

N

O

R

S

(continued)

V

W

X

GPSR Compliance

The European Union's (EU) General Product Safety Regulation (GPSR) is a set of rules that requires consumer products to be safe and our obligations to ensure this.

If you have any concerns about our products, you can contact us on ProductSafety@springernature.com

In case Publisher is established outside the EU, the EU authorized representative is:

Springer Nature Customer Service Center GmbH
Europaplatz 3
69115 Heidelberg, Germany

Batch number: 10761599

Printed by Printforce, the Netherlands